T0215075

Modern Methods in Mathematical Physics

Vladimir Ryzhov · Tatiana Fedorova ·
Kirill Safronov · Shaharin Anwar Sulaiman ·
Samsul Ariffin Abdul Karim

Modern Methods
in Mathematical Physics

Integral Equations in Wolfram Mathematica

Springer

Vladimir Ryzhov
Department of Applied Mathematics
and Mathematical Modeling
St. Petersburg State Marine Technical
University (SMTU)
Saint Petersburg, Russia

Kirill Safronov
Department of Applied Mathematics
and Mathematical Modeling
St. Petersburg State Marine Technical
University (SMTU)
Saint Petersburg, Russia

Samsul Ariffin Abdul Karim ⓘD
Department of Fundamental and Applied
Sciences
Universiti Teknologi PETRONAS
Seri Iskanda, Malaysia

Tatiana Fedorova
Department of Applied Mathematics
and Mathematical Modeling
St. Petersburg State Marine Technical
University (SMTU)
Saint Petersburg, Russia

Shaharin Anwar Sulaiman
Department of Mechanical Engineering
Universiti Teknologi PETRONAS
Seri Iskandar, Malaysia

ISBN 978-981-19-4914-2   ISBN 978-981-19-4915-9 (eBook)
https://doi.org/10.1007/978-981-19-4915-9

This Springer imprint is published by the registered company Springer Nature Singapore Pte Ltd.
The registered company address is: 152 Beach Road, #21-01/04 Gateway East, Singapore 189721,
Singapore

Introduction

This book is devoted to a one section of mathematical physics named integral equations. The solutions to many classical problems in mathematical physics lead us to integral equations. For example, in writing the boundary value problem for the free vibrations of an elastic string using the Green's function, we obtain an equivalent integral equation, which is a homogeneous linear Fredholm integral equation of the second kind. On the other hand, modeling the forced vibrations of an elastic string leads us to an inhomogeneous integral equation [1]. Thus, applications of vibration theory for different problems require knowledge of integral equations.

With the help of integral equations, the most famous problems of potential theory—the problems of Dirichlet and Neumann—were solved. The problems of the theory of elasticity and thermal conductivity also can be reduced to integral equations [1, 2, 3].

Of particular interest are symmetric integral equations, to which a separate chapter is devoted in this book. With their help, it is possible to solve special problems on the study of natural vibrations of a string and membrane, to derive the stability conditions for a compressed rod, to determine the stress and displacement fields in a half-space under pressure of an absolutely rigid body [2].

Integral equations arising in problems of mathematical physics help to solve problems from various fields of science and technology. In the modern world, they play a significant role in numerous applications in the field of elasticity, plasticity, heat and mass transfer, vibration theory, hydrodynamics, filtration theory, electrostatics, electrodynamics, biomechanics, queuing theory, electrical engineering, economics, medicine and many other applied fields.

Of specific interest are exact analytical or closed solutions of integral equations. They play an important role in the correct qualitative understanding of the properties of many phenomena and processes in various fields of natural science. The fact is that in computer modeling of physical, chemical, biological, or other processes, the ability to set input parameters in the form of functions or simply numerical values plays a special role. These parameters are not fixed. They are included in the basic equations of the processes and must change when performing multiple numerical experiments (simulations). In the case of a numerical solution of the equation, the subsequent change in the input parameters with a change in the experimental conditions turns out to be difficult. In the case of integral equations, such parameters are the kernel and the free term of the equation.

It is advisable to choose the structure of the input functions so that in the future it would be easier to analyze and solve the equation. In particular, as a possible criterion for the choice of input parameters, one can accept the requirement that the integral equation of the created model admits a solution in a closed form. In this book, an attempt of such analysis is made; many solved examples are accompanied by detailed comments on the possibility of constructing an analytical solution depending on the form of the kernel and the free term of the integral equation. *Wolfram Mathematica* package [4] is chosen in this book as a tool that allows both solution of a fairly wide class of integral equations and analysis of the resulting solutions.

Wolfram Mathematica is a Wolfram Research Inc. original product, primarily aimed at technical computing for research and education. Not only software for complex mathematical calculations, *Wolfram Mathematica* is also an environment that is convenient for modeling and simulation, visualization and documentation. Most important for usability is the support for free-form language input, which allows users to type commands in plain English and immediately get results and corresponding *Mathematica* commands for future use without the need for knowledge of the syntax.

Mathematica has a wide variety of graphics functions. The clarity of the presentation of the material is one of the necessary requirements for its correct understanding, especially if this material is presented in a scientific work or is used in educational processes. The use of interactive computer graphics in this case allows for maximum clarity, as it allows to look at the problem from different angles. Many examples that offer an original solution in the *Mathematica* environment are provided with visual graphic illustrations, which certainly makes it easier to understand the solution process.

This book pays special attention to equations of a general form that depend on arbitrary functions. Many equations contain one or several free parameters; in fact, families of integral equations are considered. If desired, readers can solve all the proposed examples on their own by changing the parameters to others. It is assumed that readers are familiar with the basics of working in the *Wolfram Mathematica* package and can easily reproduce the codes given in the book, making their own changes and improvements if necessary. The exact solutions obtained in the examples can also be used to check consistency and to estimate errors using various numerical, asymptotic and approximate methods.

Exact, approximate analytical and numerical methods for solving linear integral equations are presented in this book. Of course, one small book would not be able to acquaint readers with all the existing solution methods. However, when choosing methods, preference is given to practical points; this book primarily presents methods to effectively build a solution. For the readers' better understanding of the methods, each section is provided with examples of specific equations. The book can be used by teachers of colleges and universities as a basis for courses on integral equations and equations of mathematical physics.

For the convenience of a wide audience with different mathematical education, this book is not overloaded with theoretical materials, for which the main theorems are given without proofs. The authors also try to avoid the use of special terminologies in the text. Therefore, some methods are described in a somewhat simplified way, but at the end of each chapter, there are necessary links to books in which these methods are discussed in more detail. For some equations, only solutions of the simplest form are given. The book does not cover nonlinear, two-, three- and multidimensional integral equations.

The authors hope that this book will be useful to a wide range of researchers, college and university teachers, engineers and students in various fields of mathematics, mechanics, physics, chemistry, biology, economics and engineering sciences.

The structure of presentations of the materials allows the book to be used as an online massive open online course (MOOC). The relevance of MOOC and the use of distance learning in conditions of limited mobility (such as the 2020–2021 COVID pandemic period) has been confirmed in practice. In carrying out educational processes, effective tools have become vital, as they allow remote provision of both the actual study of the educational material and the process of quality control of the acquired knowledge [5, 6].

In the case of studying engineering disciplines associated with the need for active development of practical competencies, the methodological capabilities of information and communication tools become decisive. Experience shows [7] that virtual practical works supported by e-learning systems and web conferencing systems can be such distance learning tools.

Recent studies in [5, 6] indicate that the use of computer technologies, in particular virtual practical work in any computer simulation package, especially in *Wolfram Mathematica* package, makes educational process more dynamic, flexible and compact. Of course, this is also due to the fact that the extraordinary interest and commitment of the modern young generation (new generation) to digital tools and applications plays a positive role.

Experience shows that classes conducted using distance learning technologies and computer modeling technologies generate more interest and response from students than those of traditional lectures, seminars and laboratory work. According to some experts, the use of information and communication tools can increase the efficiency of studying the disciplines of the natural science cycle by 20–30% [5, 6].

From a methodological point of view, it is important that the use of distance learning elements in educational process makes it possible to provide:

- individual training;
- the possibility of modular execution of each task;
- step-by-step study of a physical, technical or other process described by this equation;
- the possibility of meaningful analysis of the obtained numerical data and simultaneously with the simulation of the process;
- the possibility of creative modification and improvement, making adjustments to the developed (or proposed) mathematical and computer model.

This textbook is closely related to the specified topic—the use of virtual practical tasks technologies together with the capabilities of e-learning platforms to support the distance educational process.

This textbook was prepared by a team of authors from two universities: St. Petersburg State Marine Technical University (Russia) and Universiti Teknologi PETRONAS (Malaysia). The textbook demonstrates the use of applied packages, using the example of *Wolfram Mathematica* and distance learning technology previously developed by the authors within the framework of the InMotion project: Innovative teaching and learning strategies in open modeling and simulation environment for student-centered engineering education (ERASMUS+, CBHE, No. 573751-EPP-1-2016-1-DE-EPPKA2-CBHE-JP) [7].

Step-by-step instructions for solving a large number of problems related to solving integral equations, both analytically and in the *Wolfram Mathematica* environment, are designed to build models of complex systems and to study the properties of these systems within a virtual training laboratory, which can be created, for example, on based on the Sakai Learning Management System [8].

References

1. W. V. Lovitt, *Linear Integral Equations*, (Dover Publ., New York, 1950)
2. P. P. Zabreyko, A. I. Koshelev, et al., Integral Equations: A Reference Text, (Noordhoff Int. Publ., Leyden, 1975)
3. S. G. Mikhlin, *Linear Integral Equations*, (Courier Dover Publications, 2020)
4. Wolfram Mathematica, http://www.wolfram.com/mathematica/
5. I.V. Novopashenny,V.A. Ryzov, Yu.B. Senichenkov, Yu.V. Shornikov, *Innovative teaching and learning strategies in open modelling and simulation environment for student-centered engineering*: Computer Tools in Education, 2016), № 2, pp. 62–64
6. V.A. Ryzhov, Yu.V. Shornikov, Y.B Senichenkov, B.V. Sokolov et al., in "Innovative Teaching and Learning Strategies in Open Modelling and Simulation Environment for Student-centered Engineering Education" (plenary paper): Proceedings of the VIII All-Russia Scientific-Practical Conference on Simulation and its Application in Science and Industry. Simulation: The Theory and Practice, («IMMOD-2017»), October 18–20, 2017, pp. 37–45
7. InMotion, Innovative teaching and learning strategies in open modelling and simulation environment for student-centered engineering education, ERASMUS+ № 573751-EPP-1-2016-1-DE-EPPKA2-CBHE-JP, Technical Report 2019
8. Sakai Learning Management System, https://www.sakailms.org/

Contents

Chapter 1
Fundamentals. Classification of Integral Equations

1.1 Basic Types of Integral Equations: A Solution of Integral Equation

An integral equation is an equation that contains an unknown function under an integral sign.

Generally, linear integral equations can be classified as:

- Fredholm equation of the first kind

$$\int_a^b K(x,t)\varphi(t)\mathrm{d}t = f(x);$$

- Fredholm equation of the second kind

$$\varphi(x) = \lambda \int_a^b K(x,t)\varphi(t)\mathrm{d}t + f(x);$$

- Volterra equation of the first kind

$$\int_a^x K(x,t)\varphi(t)\mathrm{d}t = f(x);$$

- Volterra equation of the second kind

$$\varphi(x) = \lambda \int_a^x K(x,t)\varphi(t)\mathrm{d}t + f(x).$$

© The Author(s), under exclusive license to Springer Nature Singapore Pte Ltd. 2022
V. Ryzhov et al., *Modern Methods in Mathematical Physics*,
https://doi.org/10.1007/978-981-19-4915-9_1

For example, for $\lambda = 1$, the Fredholm equation of the second kind has the form:

$$\varphi(x) = \int_a^b K(x, t)\varphi(t)dt + f(x), \tag{1.1}$$

where $K(x, t)$ and $f(x)$ are known as functions and $\varphi(x)$ is a required function to be solved for. The function $K(x, t)$ is called the kernel of the integral equation, while the function $f(x)$ is called the free term of this equation. Variables x, t take values from some fixed segment $[a, b]$. In this case, it is assumed that the unknown function $\varphi(x)$ depends on the real variable x, which changes in the same segment $[a, b]$ as the variable of integration t. The interval $[a, b]$ can be either finite or infinite. The functions $K(x, t)$ and $f(x)$ are assumed to be known and defined almost everywhere in the square $a \leq x, t \leq b$ and in the interval $a \leq x \leq b$, respectively. A characteristic feature of this equation is its linearity, as the unknown function $\varphi(x)$ enters it linearly. But a number of problems lead us to nonlinear integral equations, for example, to the following equations:

$$\varphi(x) = \int_a^b K(x, t)g(\varphi(t))dt, \tag{1.2}$$

where $K(x, t)$ and $g(z)$ are known functions and $\varphi(x)$ is the required function. In this tutorial, we will restrict ourselves to considering only linear integral equations, as in [[1–6].

If the free term of the integral equation $f(x)$ is equal to zero, then the integral equation is called *homogeneous*; otherwise, the equation is called *inhomogeneous*.

Any function that turns the equation into an identity for any $x \in [a, b]$ upon substitution is called a *solution* of the integral equation.

This concept can be explored with examples. Below, each new concept will be illustrated with examples. Some examples used in this book are taken from the manual [7], and others are compiled by the authors themselves.

Example 1.1 Check that the function $\varphi(x) = 2e^x$ is a solution of the equation

$$\varphi(x) = \int_0^1 e^{x-t}\varphi(t)dt.$$

Solution

The given equation is a homogeneous Fredholm equation of the second kind. We substitute the function $\varphi(x)$ into the right-hand side of the equation and calculate the integral

$$\int\limits_0^1 e^{x-t}\varphi(t)dt = \int\limits_0^1 e^{x-t}2e^t\,dt = 2e^x\int\limits_0^1 dt = 2e^x.$$

Substituting $\varphi(x)$ in the left-hand side of the equation, we obtain the identity $2e^x = 2e^x$; therefore, $\varphi(x) = 2e^x$ is a solution of the equation.

To solve problems of this type, we write the function **IntEqQ** in the *Wolfram Mathematica* package [8], which returns **True** if the equation is correct, and **False** otherwise:

```
In [1]:= Clear[IntEqQ]
         IntEqQ[Kernel_,f_,φ_,λ_,{a_,b_},x_]:=
             φ[x]==λ Integrate[Kernel[x,t]φ[t],{t,a,b}]+f[x]
```

Note that this function is suitable for checking solutions not only in the case of the Fredholm equation but also in the case of the Volterra equation. The initial data are the kernel, the free term and the solution function which are specified in the form of pure functions, as well as the parameter λ and the limits of integration (which depend on the type of equation). Let us write down the initial data for this example:

```
In [3]:= Clear[Kernel,f,φ,λ]
         Kernel=Function[{x,t},Exp[x-t]];
         f=Function[{x},0];
         φ=Function[{x},2 Exp[x]];
         λ=1;
```

Substituting the initial data into the function **IntEqQ**, we get:

```
In [8]:= IntEqQ[Kernel,f,φ,λ,{0,1},x]
Out[8]:= True
```

Answer

This means that the proposed function is a solution of the integral equation.

Example 1.2 Check that the function $\varphi(x) = \frac{1}{(1+x^2)^{3/2}}$ is a solution of the Volterra integral equation.

$$\varphi(x) = \frac{1}{1+x^2} - \int\limits_0^x \frac{t}{1+x^2}\varphi(t)dt.$$

Solution

To solve the problem, we will use the function **IntEqQ** written in Example 1.1. Let us set the initial data:

```
In [9]:= Clear[Kernel,f,φ,λ]
         Kernel=Function[{x,t},t/(1+x^2)];
         f=Function[{x},1/(1+x^2)];
         φ=Function[{x},1/(1+x^2)^(3/2)];
         λ=-1;
```

Substituting the initial data into the function **IntEqQ**, we get:

```
In [14]:= IntEqQ[Kernel,f,φ,λ,{0,x},x]
Out[14]:= True
```

Answer

This means that the proposed function is a solution to the integral equation.

Example 1.3 Consider a more complex problem. Check that the function $\varphi(x) = \sin\frac{\pi x}{2}$ is a solution of the Fredholm integral equation

$$\varphi(x) = \frac{x}{2} + \frac{\pi^2}{4} \int\limits_0^1 K(x,t)\varphi(t)\mathrm{d}t,$$

where the kernel has the form

$$K(x,t) = \begin{cases} \frac{x(2-t)}{2}, & 0 \le x \le t \\ \frac{t(2-x)}{2}, & t < x \le 1 \end{cases}.$$

Solution

In this example, the function **IntEqQ,** written in Example 1.1, can be used. We set the initial data:

```
In [15]:= Clear[Kernel,f,φ,λ]
          Kernel=Function[{x,t},Piecewise[{{x(2-
t)/2,0<x<=t},
                {t(2-x)/2,t<x<1}}]];
          f=Function[{x},x/2];
```

```
φ=Function[{x},Sin[π x/2]];
λ=π^2/4;
```

Substituting data into a function **IntEqQ**

```
In [20]:= IntEqQ[Kernel,f,φ,λ,{0,x},x]
```

we get the following result:

$$\text{Out[20]:=} \mathbf{Sin}\left[\tfrac{\pi x}{2}\right] == \tfrac{x}{2} + \tfrac{1}{4}\pi^2\left(\left\{\begin{array}{ll} -\dfrac{2\left(x-2\mathrm{Sin}\left[\frac{\pi x}{2}\right]\right)}{\pi^2} & 0 < x < 1 \\ \dfrac{(-2+x)\left(\pi x\mathrm{Cos}\left[\frac{\pi x}{2}\right]-2\mathrm{Sin}\left[\frac{\pi x}{2}\right]\right)}{\pi^2} & x == 1 \\ 0 & \mathbf{True} \end{array}\right.\right)$$

In this case, we can separate the required part of the piecewise function using the function **Refine**:

```
In [21]:= Refine[%,0<x<1]
```

Then, as a result, we get:

```
Out [21]:= True
```

Answer
This means that the proposed function is a solution of the integral equation.

1.1.1 Fredholm Equation of the Second Kind

Consider the Eq. (1.1). If the kernel $K(x, t)$ satisfies the inequality

$$\left\| K(x, t)^2 \right\| = \int_a^b \int_a^b K^2(x, t)\mathrm{d}x\mathrm{d}t < +\infty, \tag{1.3}$$

then it is square-integrable in the square $[a, b] \times [a, b]$ and, therefore, belongs to the space of square-integrable functions which is usually denoted by $L_2[a, b] \times [a, b]$. Such kernels are called *Fredholm kernels* or *Hilbert–Schmidt kernels*. In addition, we will assume that a similar inequality is also satisfied for the free term

$$\| f(x)^2 \| = \int\limits_a^b f^2(x) dx < +\infty. \tag{1.4}$$

This inequality means that the function $f(x)$ is square integrable on the interval $[a, b]$ and, therefore, belongs to the $L_2[a, b]$ space. Solving Eq. (1.1), we will always assume that the unknown function $\varphi(x)$ is also square-integrable on the interval $[a, b]$ and, therefore, belongs to the $L_2[a, b]$ space.

Let the kernel $K(x, t)$ satisfy inequality (1.3). Then, the expression

$$\psi(x) = \left(\widehat{K} \varphi \right)(x) = \int\limits_a^b K(x, t) \varphi(t) dt \tag{1.5}$$

defines an integral operator acting in the $L_2[a, b]$ space. Such an operator is called the Fredholm or Hilbert–Schmidt operator. The study of Eq. (1.1) is reduced to the study of the properties of this operator.

Note. Let $\widehat{K_1}$ and $\widehat{K_2}$ are two integral operators of the form (1.5) and $K_1(x, t)$, $K_2(x, t)$—their corresponding kernels. If the operators $\widehat{K_1}$ and $\widehat{K_2}$ are equal, that is,

$$\widehat{K_1} \varphi = \widehat{K_2} \varphi$$

for all φ, from the space $L_2[a, b]$, then their kernels are equal almost everywhere

$$K_1(x, t) = K_2(x, t).$$

Indeed, if

$$\widehat{K_1} \varphi - \widehat{K_2} \varphi = \int_a^b (K_1(x, t) - K_2(x, t)) \varphi(t) dt = 0$$

for all φ from the $L_2[a, b]$ space, then for almost all x from the interval $[a, b]$,

$$\int\limits_a^b |K_1(x, t) - K_2(x, t)|^2 \, dt = 0$$

and then

$$\int\limits_a^b \int\limits_a^b |K_1(x, t) - K_2(x, t)|^2 dx dt = 0,$$

whence our statement follows. Thus, if we do not distinguish equivalent summable functions, then we can say that the correspondence between integral operators and kernels is one-to-one.

1.1.2 Fredholm Equation of the First Kind

The Fredholm equation of the first kind is an integral equation in which the unknown function is present only under the integral

$$\int\limits_a^b K(x, t)\varphi(t)dt = f(x). \tag{1.6}$$

The kernel $K(x, t)$ and the free term $f(x)$ of this equation also satisfy conditions (1.3) and (1.4).

1.1.3 Volterra Equation of the Second Kind

The Volterra equation of the second kind has the form

$$\varphi(x) = \int\limits_a^x K(x, t)\varphi(t)dt + f(x), \tag{1.7}$$

The variable x runs over some fixed interval $[a, b]$. In particular, the cases $a = -\infty$, $b = \infty$ are possible.

Speaking about the Volterra equation, in the general case, we will not formulate the restrictions imposed on the kernel and the free term of the equation.

Under some restrictions, the Volterra equation can be considered as a special case of the Fredholm equation. Let $f(x)$ satisfy inequality (1.4), and let the kernel $K(x, t)$ satisfy the inequality

$$\int\limits_a^b \left\{ \int\limits_a^x |K(x, t)|^2 dt \right\} dx < \infty.$$

The kernel $K(x, t)$ within the meaning of the problem is defined on the interval $a \leq t \leq x$; let us extend it for values $t > x$, by setting

$$K(x, t) = 0, x < t \leq b.$$

Then, Eq. (1.7) can be written in the form (1.1), and the kernel satisfies inequality (1.3).

1.1.4 Volterra Equation of the First Kind

The Volterra equation of the first kind is characterized by the absence of the term $\varphi(x)$ outside the integral. It has the form

$$\int_a^x K(x, t)\varphi(t)\mathrm{d}t = f(x). \tag{1.8}$$

1.2 Equations with a Weak Singularity

An equation with a weak singularity is an integral equation in which the kernel has the form

$$K(x, t) = \frac{A(x,t)}{|x-t|^\alpha}, 0 < \alpha < 1, \tag{1.9}$$

where $A(x, t)$ is a continuous function on the square $a \leq x, t \leq b$.

The integral operator (1.5), defined by the kernel with a weak singularity (1.9), is called an integral operator with a weak singularity.

In the case $\alpha < \frac{1}{2}$, we deal with a Fredholm kernel.

Example 1.4 The equation is given

$$\varphi(x) - \lambda \int_0^x \frac{\varphi(t)}{(x - t)^\alpha}\mathrm{d}t = f(x), 0 < \alpha < \frac{1}{2}.$$

Determine the type of the equation and check that the kernel is of a Fredholm type.

Solution

The given equation is an inhomogeneous Volterra equation of the second kind. Kernel

$$K(x, t) = \frac{1}{(x - t)^\alpha}$$

has a discontinuity at $t = x$, and thus, this kernel has a weak singularity.

Taking $K(x, t) \equiv 0$ for $x < t \leq$ we find the norm of the kernel:

$$\|K(x, t)\|^2 = \int_a^b \int_a^b K^2(x, t)dxdt = \int_0^a dx \int_0^x \frac{dt}{(x - t)^{2\alpha}}$$

$$= \int_0^a \frac{x^{-2\alpha+1}}{-2\alpha + 1}dx = \frac{a^{2-2\alpha}}{(1 - 2\alpha)(2 - 2\alpha)} < +\infty.$$

Answer

So, the kernel of the equation is of a Fredholm type.

Volterra integral equations with a weak singularity can be either of the first or of the second kind. Let us give an example of the Volterra equation of the second kind.

Example 1.5 Solve Volterra integral equation with weak singularity

$$\varphi(x) = x^c - \int_0^x \frac{\varphi(t)}{\sqrt{x - t}}dt, \quad 0 \leq x \leq 1, \quad c > 0$$

Solution

Let us set the given equation in the *Wolfram Mathematica* package as an "equation".

```
In [1]:= Clear[eqn,sol,φ,x,c]
         eqn=φ[x]==x^c-Integrate[φ[t]/Sqrt[x-t],{t,0,x}];
         sol=DSolveValue[eqn, φ[x],{x,0,1}]
```

The solution of this equation is

$$Out[3]:= x^c + \frac{e^{\pi x}\pi^{-c}\text{Gamma}[1+c]\text{Gamma}[\frac{1}{2}+c,\pi x]}{\text{Gamma}[\frac{1}{2}+c]} - e^{\pi x}\pi^{-c}\textbf{Gamma}[1+c, \pi x]$$

Plotting the graphs of the solution for different values of the parameter c

```
In [4]:= optplot={AxesStyle->Arrowheads[{0.0,0.025}],
            AxesLabel->{Style[x,Bold,Medium],
         Style[y,Bold,Medium]}};
            Plot[Evaluate[Table[sol,{c,0.5,3,0.5}]],{x,0,1},
            Evaluate@optplot,
            PlotLegends->ToString/@Thread[c==Range[0,3,0.5]]]
```

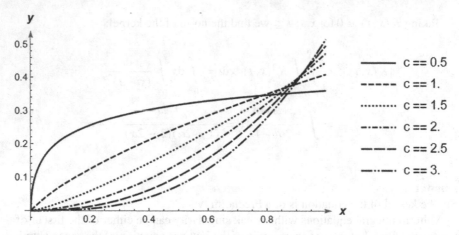

Fig. 1.1 Family of solutions of the Volterra equation with a weak singularity for different values of the parameter c (Example 1.5)

as a result, we have a family of curves, Fig. 1.1.

Answer

$$\varphi(x) = x^c + \frac{e^{\pi x} \pi^{-c} \text{Gamma}\left[1 + c\right] \text{Gamma}\left[\frac{1}{2} + c, \pi x\right]}{\text{Gamma}\left[\frac{1}{2} + c\right]}$$
$$- e^{\pi x} \pi^{-c} \text{Gamma}[1 + c, \pi x]$$

An example of a Volterra equation of the first kind with a kernel with weak singularity is the Abel integral equation, which will be considered in Sect. 1.3.

1.3 Abel Problem: Abel Integral Equation

Historically, the Abel problem is one of the first problems that led to the need to consider integral equations. In 1823, Niels Henrik Abel generalized the tautochron problem. Let us formulate this problem.

A point particle moves under gravity in the vertical plane (ξ, η) along a certain curve l. It is required to define the curve l so that the particle, having started its motion without initial velocity at a point of the curve with the ordinate y, reaches the ξ axis in time $t = f_1(y)$, where $f_1(y)$ is a given function depending on the initial height y. (see Fig. 1.2). Note that in the classical tautochron problem, $f_1(y) = \text{const}$; i.e., the time to reach the ξ axis does not depend on the initial position of the particle.

The absolute value of the velocity of a moving point is determined by the acceleration g along the η axis

$$v = \sqrt{2g(y - \eta)}.$$

Fig. 1.2 Material point
motion in the Abel problem

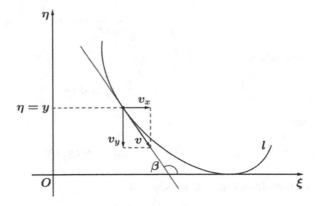

Denote by $= (\eta)$. an angle of inclination of the tangent to the axis ξ (Fig. 1.2).
Then, for the velocity component along the η. axis, we have

$$v_\eta = \frac{d\eta}{dt} = -\sqrt{2g(y-\eta)} \sin\beta,$$

and

$$dt = -\frac{d\eta}{\sqrt{2g(y-\eta)} \sin\beta}.$$

We do not know the shape of the curve l, so we put

$$\frac{1}{\sin\beta} = \frac{1}{\sin\beta(\eta)} = \varphi(\eta)$$

and integrate over η in the range from 0 to y. We obtain the *Abel integral equation*:

$$-\sqrt{2g}\,f_1(y) = \int_0^y \frac{\varphi(\eta)d\eta}{\sqrt{y-\eta}},$$

Denote $-\sqrt{2g}\,f_1(y)$ by $f(y)$, and finally get:

$$f(y) = \int_0^y \frac{\varphi(\eta)d\eta}{\sqrt{y-\eta}}, \tag{1.10}$$

where $\varphi(\eta)$ is the required function, $f(y)$ is the given function. After finding $\varphi(\eta)$.
from this integral equation, we can formulate the equation for the required curve l.
Indeed,

$$\frac{1}{\sin \beta} = \varphi(\eta),$$

from where

$$\eta = \Phi(\beta)$$

and

$$d\eta = \Phi'(\beta)d\beta.$$

further, from Fig. 1.2, we see that

$$\frac{d\eta}{d\xi} = tg\beta,$$

from where

$$d\xi = \frac{\Phi'(\beta)d\beta}{tg\beta}$$

and

$$\xi = \int \frac{\Phi'(\beta)d\beta}{tg\beta} = \Phi_1(\beta).$$

Thus, the required curve l is determined by the parametric equations

$$l : \begin{cases} \xi = \Phi_1(\beta) \\ \eta = \Phi(\beta) \end{cases}$$

So, Abel's problem is reduced to solving Eq. (1.10), which is the Volterra equation of the first kind. We also note that in Eq. (1.10) the kernel has a weak singularity.

Consider now a more general case of the Abel integral equation

$$f(x) = \int_0^x \frac{\varphi(t)}{(x-t)^\alpha}dt, \quad 0 < \alpha < 1. \tag{1.11}$$

We will assume that the function $f(x)$ is continuously differentiable on some interval $[0, a]$. This equation admits an inversion formula that can be easily derived as follows.

Replace in Eq. (1.11) x by s, multiply both sides of the equation by $ds/(x-s)^{1-\alpha}$ and integrate over s from 0 to x. We get

$$\int_0^x \frac{ds}{(x-s)^{1-\alpha}} \int_0^s \frac{\varphi(t)dt}{(s-t)^{\alpha}} = F(x),$$

where

$$F(x) = \int_0^x \frac{f(s)ds}{(x-s)^{1-\alpha}}.$$

Changing on the left the order of integration

$$\int_0^x \varphi(t)dt \int_t^x \frac{ds}{(x-s)^{1-\alpha}(s-t)^{\alpha}} = F(x) \tag{1.12}$$

and making the substitution $s = t + y(x - t)$ after elementary transformations, we get

$$\int_t^x \frac{ds}{(x-s)^{1-\alpha}(s-t)^{\alpha}} = \int_0^1 \frac{dy}{y^{\alpha}(1-y)^{1-\alpha}} = \frac{\pi}{\sin \alpha \pi}.$$

Substituting the obtained expression in (1.12), we obtain the required inversion formula

$$\varphi(x) = \frac{\sin \alpha \pi}{\pi} F'(x),$$

or substituting the expression for the function $F(x)$,

$$\varphi(x) = \frac{\sin \alpha \pi}{\pi} \left(\frac{f(0)}{x^{1-\alpha}} + \int_0^x \frac{f'(s)}{(x-s)^{1-\alpha}} ds \right). \tag{1.13}$$

Example 1.6 Analyze the solution of Eq. (1.10) in the *Wolfram Mathematica* package, starting with the case $f(y) = c$, i.e., with the classical tautochron problem.

Solution
The solution cannot be found by the approach demonstrated in Example 1.5.

```
In [1]:= Clear[eqn, sol,f,φ,η,y]
         f[y]=c;
         eqn=f[y]==Integrate[φ[η]/Sqrt[y-η],{η,0,y}];
         sol=DSolveValue[eqn,φ[η],η]
```

$$\text{Out[4]:}= \textbf{DSolveValue}\left[c == \int_0^y \frac{\phi[\eta]}{\sqrt{y-\eta}}d\eta,\ \phi[\eta],\ \eta\right]$$

Therefore, we will use formula (1.13) to construct an analytical solution.

```
In [5]:= Clear[f,φ,η,y,c,α]
         f=Function[{y},c];
         α=1/2;
         φ[η]=Sin[α π]/π(f[0]/η^(1-α)+
           Integrate[f'[s]/(η-s)^(1-α),{s,0,η}])
```
$$\text{Out[8]:}= \frac{c}{\pi\sqrt{\eta}}$$

Let us make sure that the resulting function is a solution of Eq. (1.10).

```
In [9]:= f[y]==Integrate[φ[η]/Sqrt[y-η],{η,0,y}]
```
$$\text{Out[9]:}= \textbf{True if Re}[y]> \textbf{0 \&\&Im}[y]== \textbf{0}$$

Since the physical meaning of y is the height above some zero level, the function $\varphi(\eta)$ is in fact the solution.

Next, we get the parametric equation of the desired curve.

```
In [10]:= y[β_]:=SolveValues[1/Sin[β]==φ[η],η][[1]]
          x[β_]:=Integrate[y'[β]/Tan[β],β]
```

Let us display the solution:

```
In [12]:= x[β]
          y[β]
```
$$\text{Out[12]:}= \frac{2c^2\left(\frac{\beta}{2} + \frac{1}{4}\text{Sin}[2\beta]\right)}{\pi^2}$$

$$\text{Out[13]:}= \frac{c^2\text{Sin}[\beta]^2}{\pi^2}$$

This is the parametric equation of the cycloid. To make the result more visual, we will plot a graph of the obtained solution for $c = 1$ (Fig. 1.3). This is how the material point will move.

```
In [14]:= optplot={AxesStyle->Arrowheads[{0.0,0.025}],
          AxesLabel->{Style[x,Bold,Me-
dium],Style[y,Bold,Medium]}};
          ParametricPlot[Evaluate[{x[β],y[β]}/.c-
>1],{β,π/2,π},
          Evaluate@optplot]
```

Let us increase the considered interval and construct a family of cycloids with different values of the parameter c (Fig. 1.4).

```
In [16]:= Clear[c]
          ParametricPlot[Evaluate[Ta-
ble[{x[β],y[β]},{c,1,3,0.5}]],
```

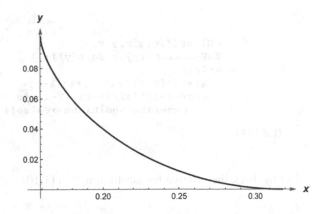

Fig. 1.3 Motion trajectory of the material point in the Abel problem

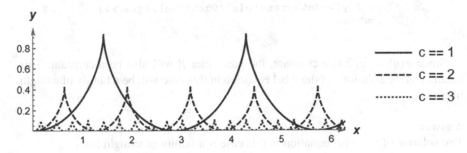

Fig. 1.4 Family of cycloids in the Abel problem

```
            {β,0,20π},AspectRatio->1/3,PlotRange-
    >1/3,PlotRange->{{0,2π},{0,1}},
            Evaluate@optplot,PlotLegends->
                Placed[ToString/@Thread[c==Range[1,3,0.5]],Af-
    ter]]
```

Answer

The solution to the Abel problem is a cycloid defined by a parametric function

$$\begin{cases} x(\beta) = \frac{2c^2\left(\frac{\beta}{2}+\frac{1}{4}\sin 2\beta\right)}{\pi^2} \\ y(\beta) = \frac{c^2 \sin \beta}{\pi^2} \end{cases}.$$

Example 1.7 Obtain a solution of the Abel Eq. (1.10) in the case $f(y) = c\sqrt{y}$.

Solution

We can use the same approach as in the previous example. We will also add the option
GenerateConditions→False to obtain the result without additional assumptions.

```
In [18]:= Clear[f,φ,η,x,y,c,α]
          f=Function[{y},c Sqrt[y]];
          α=1/2;
          φ[η]=Sin[α π]/π(f[0]/η^(1-α)+
          Integrate[f'[s]/(η-s)^(1-α),{s,0,η},
              GenerateConditions->False])
Out[21]:= c/2
```

Let us check the solution by substituting it in (1.10).

```
In [22]:= f[y]==Integrate[φ[η]/Sqrt[y-η],{η,0,y}]
Out[22]:= True
```

Since $\varphi(\eta) = c/2$ is a constant, the parameter β will also be a constant. This
means that the solutions of the Abel equation in this case will be a family of straight
lines.

Answer

The solution of the Abel equation in this case is a family of straight lines.

1.4 Solution of Integral Equations by the Differentiation Method

Volterra equations of the first and second kind can often be solved by reducing to the Cauchy problem for a differential equation which is obtained by differentiating the given integral equation. Consider the Volterra equation of the second kind

$$\varphi(x) = \lambda \int\limits_{\alpha}^{x} K(x,t)\varphi(t)\mathrm{d}t + f(x). \qquad (1.14)$$

Let us differentiate Eq. (1.14) with respect to x using the rule for differentiating the integral depending on the parameter

$$\varphi'(x) = \lambda \int\limits_{\alpha}^{x} \tfrac{\partial K(x,t)}{\partial x}\varphi(t)\mathrm{d}t + \lambda K(x,x)\varphi(x) + f'(x). \qquad (1.15)$$

If for a given kernel the obtained integral can be expressed in terms of the initial one, then we find a differential equation of the first order with respect to the unknown function $\varphi(x)$ by expressing the integral from (1.14) and substituting it into (1.15). The initial condition to the Cauchy problem can be found from (1.14), substituting $x = a$, i.e., $\varphi(\alpha) = f(\alpha)$, so the solution of Eq. (1.14) is reduced to the solution of the Cauchy problem.

Sometimes, twice differentiating of the given equation may help to get rid of the integral. In this case, the initial conditions are determined from the given Eq. (1.14) and the differentiated one (1.15).

The differentiation method is applicable for a fairly limited class of Volterra integral equations. As a rule, kernels of such equations either represent the sum of exponential or trigonometric functions or are polynomials of the variable x.

The differentiation method can also be used to solve the integral equations of Volterra of the first kind. In this case, as a result of differentiation, an integral Volterra equation of the second kind is obtained, which can be solved either by repeated differentiation or by some other method.

Some Fredholm equations can also be solved by the differentiation method. Sometimes, this requires breaking up the region of integration.

Example 1.8 Solve the Volterra equation using the differentiation method

$$\varphi(x) = x - \int\limits_{0}^{x} e^{x-t}\varphi(t)\mathrm{d}t.$$

Solution
This equation is Volterra equation of the second kind. Differentiate it with respect to x.

$$\varphi'(x) = 1 - \int_0^x e^{x-t}\varphi(t)dt - e^{x-x}\varphi(x).$$

The remaining integral can be replaced by the expression from the original equation

$$\int_0^x e^{x-t}\varphi(t)dt = -\varphi(x) + x.$$

Then, the differentiated equation takes the form

$$\varphi'(x) = 1 - \varphi(x) + \varphi(x) - x,$$

or

$$\varphi(x) = 1 - x.$$

The initial condition is determined from the given equation, setting in it $x = 0$:

$$\varphi'(x) = 0.$$

So, the solution of the Volterra equation has been reduced to the Cauchy problem:

$$\varphi'(x) = 1 - x;$$
$$\varphi(0) = 0.$$

General solution of the equation

$$\varphi(x) = x - \frac{x^2}{2} + C,$$

where C is arbitrary constant. From the initial condition, we have $C = 0$. The solution of the Cauchy problem and the unique solution of an integral equation is

$$\varphi(x) = x - \frac{x^2}{2}.$$

Now let's implement the differentiation method in *Wolfram Mathematica*. Since the idea of the method is to elinate the integral from the original equation and the differentiated one, we can use the capabilities of *Wolfram Mathematica* when working with symbolic computations.

Set the initial data in the form of an equation:

```
In [1]:= Clear[Kernel,f,φ,x]
         Kernel:=Function[{x,t},Exp[x-t]]
         f:=Function[{x},x]
         λ=-1;
         inteqn=φ[x]==λ Integrate[Ker-
nel[x,t]φ[t],{t,0,x}]+f[x]
```
$$\text{Out[5]:}= \phi[x] == x - \int_0^x e^{-t+x}\phi[t]dt$$

Then, differentiate the equation.

```
In [6]:= D[inteqn,x]
```
$$\text{Out[6]:}= \phi'[x] == 1 - \int_0^x e^{-t+x}\phi[t]dt - \phi[x]$$

The integrals in the equations coincide. We can do it in the following way

```
In [7]:= difeqn=Eliminate[{inteqn,D[inteqn,x]},
         Integrate[E^(-t+x)*φ[t],{t,0,x}]]
```
$$\text{Out[7]:}= 1 - \phi'[x] == x$$

We can find the initial conditions, setting $x = 0$ in the original equation.

```
In [8]:= initcond=inteqn/.x->0
```
$$\text{Out[8]:}= \phi[0] == 0$$

Now, it remains to solve the Cauchy problem:

```
In [9]:= sol=DSolve[{difeqn,initcond},φ,x][[1,1]]
```
$$\text{Out[9]:}= \phi \to \text{Function}\left[\{x\}, \tfrac{1}{2}\left(2x - x^2\right)\right]$$

The result is a pure function. One can see that the solution coincided with the manual method. Let us check that this function is a solution by substituting it into the original equation.

```
In [10]:= inteqn/.sol//FullSimplify
Out[10]:= True
```

Answer

$$\varphi(x) = x - \frac{x^2}{2}.$$

Example 1.9 Solve the Volterra equation using the differentiation method

$$\int\limits_0^x (1 - x^2 + t^2)\varphi(t)dt = \frac{x^2}{2}.$$

Solution
This is Volterra integral equation of the first kind. Let us differentiate it by x

$$-2x \int\limits_0^x \varphi(t)dt + \varphi(x) = x,$$

whence

$$\varphi(x) = x + 2x \int\limits_0^x \varphi(t)dt.$$

This is the Volterra integral equation of the second kind for which the differentiation method can be used

$$\varphi'(x) = 1 + 2 \int\limits_0^x \varphi(t)dt + 2x\varphi(x).$$

Eliminating the integral from the last two equalities, we obtain

$$\varphi'(x) = \left(\frac{1}{x} + 2x\right)\varphi(x).$$

This is a separable first-order differential equation. We solve it by separating the variables

$$\frac{d\varphi}{\varphi} = \left(\frac{1}{x} + 2x\right)dx;$$

$$\ln|\varphi| = \ln|x| + x^2 + \ln C;$$

the general solution is

$$\varphi(x) = Cxe^{x^2}. \tag{1.16}$$

Since the Volterra equation must have the unique solution, it is necessary to determine the value of the constant C. It is impossible to find out the initial condition from the original equation, since it turns into an identity when $x = 0$ for any function $\varphi(x)$. We can find the value of C, by substituting $\varphi(x)$ in the equation of the second kind, obtained after the first differentiation. Substituting (1.16) into an equation of the second kind, we get

$$Cxe^{x^2} = x\left(1 + 2\int\limits_{0}^{x} Cte^{t^2}dt\right)$$

Canceling $x \neq 0$ and calculating the integral, we find $C = 1$. So, we get

$$\varphi(x) = xe^{x^2}.$$

We can solve this equation in *Wolfram Mathematica*, somewhat generalizing the approach considered in the previous example. Let us set the equation.

```
In [11]:= Clear[Kernel,f,φ,x]
          Kernel:=Function[{x,t},1-x^2+t^2]
          f:=Function[{x},x^2/2]
          inteqn=Integrate[Kernel[x,t]φ[t],{t,0,x}]==f[x]
```
$$\text{Out[14]}:= \int_0^x \left(1+t^2-x^2\right)\phi[t]dt == \frac{x^2}{2}$$

In manual calculations, we took a factor independent of the variable t from the integral. Generally speaking, the built-in function **Integrate** does not perform this operation automatically, so we will write a function to do this.

```
In [15]:= Clear[PullOut]
          PullOut[s_]:=s//.
                   {Integrate[f_+g_,it:{x_Symbol,__}]
                    :>Integrate[f,it]+Integrate[g,it],
                    Integrate[c_ f_.,it:{x_Symbol,__}]
```

```
:>c Integrate[f,it]/;FreeQ[c,x]}
```

Then, we calculate the derivative for the original equation and use the function **PullOut**.

```
In [17]:= dinteqn=D[inteqn,x]//PullOut
```
$$\text{Out[17]}:= -2x \int_0^x \phi[t]dt + \phi[x] == x$$

Since the integrals in the resulting and original equations are different, we differentiate the equation again.

```
In [18]:= d2inteqn=D[dinteqn,x]//PullOut
```
$$\text{Out[18]}:= -2 \int_0^x \phi[t]dt - 2x\phi[x] + \phi'[x] == 1$$

As a result, we have the same integral, which can be eliminated. In order to automatically extract the integral from the equation, we define the following function.

```
In [19]:= Clear[ExtractIntegral]
          ExtractIntegral[eqn_]:=Extract[eqn,Position[eqn,
             Integrate[_,_]]][[1]]
```

Now, we eliminate the integral.

```
In [21]:= difeqn=Eliminate[{dinteqn,d2inteqn},Inte-
grate[φ[t],{t,0,x}]]
```
$$\text{Out[21]}:= x\phi'[x] == (1 + 2x^2)\phi[x]$$

We obtain the general solution of the differential equation in the form of a pure function (so that it can be substituted into the original equation).

```
In [22]:= gensol=DSolve[{difeqn},φ,x][[1,1]]
```
$$\text{Out[22]}:= \phi \rightarrow \textbf{Function}\left[\{x\}, e^{x^2}xc_1\right]$$

In order to find the constant c_1, we substitute the obtained solution into the original equation.

```
In [23]:= constsol=Solve[integn/.gensol, C[1]][[1,1]]
Out[23]:= c₁→ 1
```

As a result, we get the solution

```
In [24]:= sol=gensol/.constsol
Out[24]:= φ→ Function[{x}, eˣ²x]
```

Let us check the solution by substituting it into the original equation:

```
In [25]:= integn/.sol//FullSimplify
Out[25]:= True
```

Answer

$$\varphi(x) = xe^{x^2}.$$

Example 1.10 Solve the Volterra equation

$$\int_0^x 2^{x-t}\varphi(t)\mathrm{d}t = x^2.$$

Solution
Let us set the equation in *Wolfram Mathematica*

```
In [26]:= Clear[Kernel,f,φ,x]
          Kernel:=Function[{x,t},2x-t]
          f:=Function[{x},x^2]
          integn=Integrate[Kernel[x,t]φ[t],{t,0,x}]==f[x]
```

$$\text{Out[29]}:= \int_0^x 2^{-t+x}\phi[t]dt == x^2$$

Differentiate the equation.

```
In [30]:= dinteqn=D[inteqn,x]//PullOut
```
$$\text{Out[30]}:= \left(\int_0^x 2^{-t+x}\phi[t]dt\right)\log[2]+\phi[x]== 2x$$

Eliminate the integral.

```
In [31]:= difeqn=Eliminate[{inteqn,dinteqn},Ex-
tractIntegral[dinteqn]]
```
$$\text{Out[31]}:= \phi[x] == x(2-x\log[2])$$

The obtained equation does not contain a derivative what means that the expression on the right side is a solution. Let us verify this by direct substitution into the original equation. To do this, we define a solution in the form of a transformation rule.

```
In [32]:= sol=φ->Function[{x},Evaluate[difeqn[[2]]]]
```
$$\text{Out[32]}:= \phi\rightarrow \text{Function}\,[\{x\}, x(2-x\log[2])]$$

and apply it to the original equation.

```
In [33]:= inteqn/.sol//FullSimplify
Out[33]:= True
```

Answer

$$\varphi(x) = x(2 - x\ln 2).$$

Example 1.11 Solve the Fredholm equation

$$\int_0^\pi K(x, t)\varphi(t)dt = 3\sin x - \sin 3x$$

with a symmetric kernel

$$K(x, t) = \begin{cases} \frac{t(\pi-x)}{\pi}, & 0 \le t \le x \\ \frac{x(\pi-t)}{\pi}, & x \le t \le \pi \end{cases}.$$

Solution

This is a Fredholm equation of the first kind. The kernel is given by different formulas on different intervals, so we divide the integration interval $[0, \pi]$ into the intervals $[0, x]$ and $[x, \pi]$. The equation takes the form

$$\int_0^x \frac{t(\pi - x)}{\pi}\varphi(t)dt + \int_x^\pi \frac{x(\pi - t)}{\pi}\varphi(t)dt = 3\sin x - \sin 3x,$$

or

$$\frac{\pi - x}{\pi}\int_0^x t\varphi(t)dt + \frac{x}{\pi}\int_x^\pi (\pi - t)\varphi(t)dt = 3\sin x - \sin 3x.$$

Let us differentiate the equation with respect to x

$$-\frac{1}{\pi}\int_0^x t\varphi(t)dt + \frac{\pi - x}{\pi}x\varphi(x) + \frac{1}{\pi}\int_x^\pi (\pi - t)\varphi(t)dt$$

$$-\frac{x}{\pi}(\pi - x)\varphi(x) = 3\cos x - 3\cos 3x.$$

Now, we reduce the same terms and differentiate the equation again

$$-\frac{x}{\pi}\varphi(x) - \frac{\pi - x}{\pi}\varphi(x) = -3\sin x + 9\sin 3x,$$

finally we obtain

$$\varphi(x) = 3\sin x - 9\sin 3x.$$

This example can also be solved in *Wolfram Mathematica* but due to the difficulties with integrating piecewise functions, we should manually split the integral into two ones. Let us set the initial data.

```
In [34]:= Clear[Kernel1,Kernel1,f,φ,x]
         Kernel1:=Function[{x,t},t (π-x)/π]
         Kernel2:=Function[{x,t},x (π-t)/π]
         f:=Function[{x},3Sin[x]-Sin[3x]]
         inteqn=Integrate[Kernel1[x,t]φ[t],{t,0,x}]+
            Integrate[Kernel2[x,t]φ[t],{t,x,π}]==f[x]
```

$$\text{Out[38]}:= \int_0^x \frac{t(\pi-x)\phi[t]}{\pi} dt + \int_x^\pi \frac{(\pi-t)x\phi[t]}{\pi} dt == 3\text{Sin}[x] - \text{Sin}[3x]$$

Now, we differentiate the equation two times.

```
In [39]:= dinteqn=D[inteqn,x]//PullOut
         d2inteqn=D[dinteqn,x]//PullOut
```

$$\text{Out[39]}:= \frac{\int_x^\pi (\pi-t)\phi[t]dt}{\pi} - \frac{\int_0^x t\phi[t]dt}{\pi} = 3\text{Cos}[x] - 3\text{Cos}[3x]$$

$$\text{Out[40]}:= -\frac{(\pi-x)\phi[x]}{\pi} - \frac{x\phi[x]}{\pi} == -3\text{Sin}[x]+9\text{Sin}[3x]$$

And, then we get a solution.

```
In [41]:= sol=φ->Function[{x},Evalu
ate[Solve[d2inteqn,φ[x]][[1,1,2]]]]
```

$$\text{Out[41]}:= \phi\to \text{Function}[\{x\}, 3(\text{Sin}[x] - 3\text{Sin}[3x])]$$

Let us check the solution.

```
In [42]:= inteqn/.sol//FullSimplify
Out[42]:= True
```

Answer

$$\varphi(x) = 3\sin x - 9\sin 3x.$$

References

A. N. Kolmogorov, and S. V. Fomin, *Elements of the Theory of Functions and Functional Analysis*, (Dover Publications, New York, 1999)

S. G. Mikhlin, *Mathematical physics; an advanced course*, (Amsterdam, North-Holland Pub. Co., 1970) [1971]

P. P. Zabreyko, A. I. Koshelev, et al., Integral Equations: A Reference Text, (Noordhoff Int. Publ., Leyden, 1975)

W. V. Lovitt, *Linear Integral Equations*, (Dover Publ., New York, 1950)

J. A. Cochran, The Analysis of Linear Integral Equations (– New York: McGraw-Hill Book Co., 1972)

R. Gorenflo , S. Vessella *Abel Integral Equations: Analysis and Applications*. (– Berlin – New York: Springer – Verlag, 1991)

M. L. Krasnov, A. I. Kiselev, and G. I Makarenko, *Problems and Exercises in Integral Equations*, (Mir Publ., Moscow, 1971)

Wolfram Mathematica, http://www.wolfram.com/mathematica/

Chapter 2
Integral Equations with Difference Kernels

2.1 Difference Kernel Concept. Solution of Integral Equations with Difference Kernels by the Method of Differentiation

Integral equations with difference kernels are widely represented in the special scientific literature, for example, see references [1–3]. Such integral equations are also considered in different applied problems [10]. The Fredholm equation with a difference kernel has the form

$$a\varphi(x) - \int_{-\infty}^{+\infty} K(x-t)\varphi(t)\mathrm{d}t = f(x), \tag{2.1}$$

where a is a constant and the kernel $K(x-t)$ depends only on the difference of the real variables x and t. It is also called a convolution-type equation.

Sometimes, the "almost difference" equations are considered

$$a(x)\varphi(x) - \int_{-\infty}^{+\infty} K(x, x-t)\varphi(t)\mathrm{d}t = f(x),$$

in which the dependence of the kernel $K(x, x-t)$ on the first argument is insignificant in some sense.

The Volterra equations with difference kernels are of well-known interest [2, 3]; the equations are given below:

$$a\varphi(x) - \int_{-\infty}^{x} K(x-t)\varphi(t)\mathrm{d}t = f(x) \tag{2.2}$$

V. Ryzhov et al., *Modern Methods in Mathematical Physics*,
https://doi.org/10.1007/978-981-19-4915-9_2

or

$$a\varphi(x) - \int_0^x K(x - t)\varphi(t)\mathrm{d}t = f(x). \qquad (2.3)$$

Equation (2.3) is a special case of the equation with the difference kernel (2.2) and is obtained from the latter if we put $K(\tau) = 0, \tau < 0$ in it.

An equation with a difference kernel on the semiaxis is called the Wiener–Hopf equation:

$$a\varphi(x) - \int_0^{+\infty} K(x - t)\varphi(t)\mathrm{d}t = f(x), \quad a = \mathrm{const}$$

The paired equation, which is sometimes called dual, has the following form:

$$a\varphi(x) - \int_{-\infty}^{+\infty} K_1(x - t)\varphi(t)\mathrm{d}t = f(x), \quad x > 0$$

$$b\varphi(x) - \int_{-\infty}^{+\infty} K_2(x - t)\varphi(t)\mathrm{d}t = f(x), \quad x < 0$$

where a and b are constants.

To solve the difference equations, integral transformations are used: for the Volterra equation, the Laplace transform [4–6], and for the Fredholm equation with infinite limits of integration, the Fourier transform [7–9].

Example 2.1

a. Solve an integral equation

$$\varphi(x) = x - \int_0^x (x - t)\varphi(t)\mathrm{d}t.$$

Solution

The solution of this equation can be found in the *Wolfram Mathematica* [11] using the differentiation method, discussed in Sect. 1.4. Let us set the equation:

```
In [1]:= Clear[φ,x]
         inteqn=φ[x]==x-Integrate[(x-t)φ[t],{t,0,x}]
```
Out[2]:= $\phi[x] ==x- \int_0^x (-t+x)\phi[t]dt$

It is easy to see that to eliminate the integral, the equation needs to be differentiated twice.

```
In [3]:= dinteqn=D[inteqn,x]
         d2inteqn=D[dinteqn,x]
```
Out[3]:= $\phi'[x]== 1- \int_0^x \phi[t]dt$
Out[4]:= $\phi''[x]== -\phi[x]$

Since we have obtained a second-order differential equation, it is required to set two initial conditions. Let us do this and solve the equation.

```
In [5]:= initcond={inteqn,dinteqn}/.x->0
         sol=DSolve[{d2inteqn,initcond},φ,x]
```
Out[5]:= $\{\phi[0]== 0, \phi'[0]== 1\}$
Out[6]:= $\{\{\phi \rightarrow \text{Function}[\{x\}, \text{Sin}[x]]\}\}$

The solution of this integral equation can also be obtained using the built-in function **DSolve** (however, this does not always work).

```
In [7]:= DSolve[inteqn,φ,x]
```
Out[7]:= $\{\{\phi \rightarrow \text{Function}[\{x\}, \text{Sin}[x]]\}\}$

b. Consider the Volterra equation in a more general form

$$\varphi(x) = x - \lambda \int_0^x (x - t)\varphi(t)dt.$$

Solution

It can be noted that this equation differs from the previous one by the parameter λ, which we will need further when studying the properties of integral equations.

```
In [8]:= Clear[φ,x,λ]
         inteqn=φ[x]==x-λ Integrate[(x-t)φ[t],{t,0,x}]
         sol=DSolve[inteqn,φ,x]
```

$$\text{Out}[9]:= \phi[x]==x-\lambda \int_0^x (-t+x)\phi[t]dt$$

$$\text{Out}[10]:= \left\{\left\{\phi\rightarrow \textbf{Function}\left[\{x\},\frac{\text{Sin}\left[x\sqrt{\lambda}\right]}{\sqrt{\lambda}}\right]\right\}\right\}$$

Answer

a. $\varphi(x) = \sin x$

b. $\varphi(x) = \frac{\text{Sin}(x\sqrt{\lambda})}{\sqrt{\lambda}}$

Example 2.2 Solve integral equation

$$\varphi(x) = x^3 - \lambda \int\limits_0^x (x - t)\varphi(t)\mathrm{d}t.$$

Solution

This equation is similar to Example 2.1 but a little more complicated (Fig. 2.1).
  Let us get a solution using **DSolve**.

```
In [11]:= Clear[φ,x,λ]
          inteqn=φ[x]==x^3-λ Integrate[(x-t)φ[t],{t,0,x}]
          sol=DSolve[inteqn,φ,x]
```

$$\text{Out}[12]:=\phi[x]==x^3-\lambda \int_0^x (-t+x)\phi[t]dt$$

$$\text{Out}[13]:=\left\{\left\{\phi\rightarrow \textbf{Function}\left[\{x\},\frac{6x}{\lambda}-\frac{6\text{Sin}\left[x\sqrt{\lambda}\right]}{\lambda^{3/2}}\right]\right\}\right\}$$

And plot the graphs of the family of solutions (Fig. 2.1):

```
In [14]:= optplot={AxesStyle->Arrowheads[{0.0,0.025}],
          AxesLabel->{Style[x,Bold,Medium],
          Style[y,Bold,Medium]}};
          Plot[Evaluate[Table[φ[x]/.sol,{λ,1,3,0.5}]],{x,0,20},
          Evaluate@optplot,
```

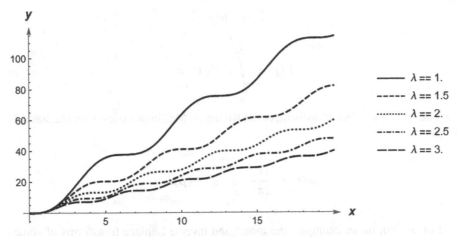

Fig. 2.1 Family of solutions of an integral equation with a difference kernel Example 2.2

```
PlotLegends->ToString/@Thread[λ==Range[1,3,0.5]]]
```

Answer

$$\varphi(x) = \frac{6x}{\lambda} - \frac{6\mathrm{Sin}\left(x\sqrt{\lambda}\right)}{\lambda^{3/2}}$$

Unfortunately, the further complication of the functions included in the integral equations leads to problems with getting an answer. The differentiation method can be used; however, it has its limitations.

Besides, for solving integral equations with difference kernels, there is such a convenient method as integral transformations [4–9].

2.2 Solution of Integral Equations and Systems of Volterra Integral Equations with Difference Kernels Using the Laplace Transform

2.2.1 Solving Volterra Integral Equations with Difference Kernels Using the Laplace Transform

To solve Volterra integral equations with difference kernels, we need such standard functions of the *Mathematica* package as the forward and inverse Laplace transforms [11]. The direct Laplace transform has the form:

$$\varphi(x) \dot= \Phi(p),$$

$$\Phi(p) = \int\limits_0^\infty e^{-pt} \varphi(t) \mathrm{d}t.$$

The inverse Laplace transform, or the Riemann–Mellin inversion formula, has the form

$$\varphi(t) = \frac{1}{2\pi i} \int\limits_{\gamma-i\infty}^{\gamma+i\infty} e^{pt} \Phi(p) \mathrm{d}p.$$

Let us find, as an example, the direct and inverse Laplace transforms of some known functions:

```
In [1]:= φ=Sin[x];
         Φ=LaplaceTransform[φ,x,p]

Out[2]:= 1/(1+p²)

In [3]:= InverseLaplaceTransform[Φ,p,x]

Out[3]:= Sin[x]
```

Now, we can solve integral equations. If $\varphi_1(x)$ and $\varphi_2(x)$ are two continuous functions defined for $x \geq 0$, then the function

$$\varphi_3(x) = (\varphi_1 * \varphi_2) = \int\limits_0^x \varphi_1(x-t)\varphi_2(t)\mathrm{d}t$$

is called the **convolution** of these two functions. The function $\varphi_3(x)$ is also defined for $x \geq 0$ and is continuous. When solving integral equations, we need an image multiplication theorem or a convolution theorem.

Theorem 2.1 (Convolution theorem) *If $\varphi_1(x)$ and $\varphi_2(x)$ are the original functions for the Laplace transform, and $\Phi_1(p)$ and $\Phi_2(p)$ are respectively their images, then the image of their convolution $\varphi_3(x)$ can be found by the formula*

$$\Phi_3(p) = \Phi_1(p) \cdot \Phi_2(p),$$

i.e., the convolution image is equal to the product of the convolutional function images.

Consider the convolution-type Volterra equation

$$\varphi(x) = \lambda \int_0^x K(x - t)\varphi(t)\mathrm{d}t + f(x). \qquad (2.4)$$

Let the functions $K(\tau)$ and $f(x)$ be continuous for $x \geqslant 0, \tau \geqslant 0$ and grow no faster than the exponential function if $x \to \infty$, so that the constants $M_1, \alpha_1, M_2, \alpha_2$ exist, so for all $\tau \geqslant 0, \ x \geqslant 0$, the inequalities hold

$$|K(\tau)| \leqslant M_1 e^{\alpha_1 \tau}; \quad |f(x)| \leqslant M_2 e^{\alpha_2 x}.$$

It can be shown that in this case, the function $\varphi(x)$ also satisfies the same condition

$$|\varphi(x)| \leqslant M e^{\alpha_3 x},$$

where

$$\alpha_3 = \max\{\alpha_1, \alpha_2\}.$$

Consequently, the Laplace image of the functions $f(x)$, $K(\tau)$ and $\varphi(x)$ can be found, and it is defined in the right half-plane $\mathrm{Re}(p) > \max\{\alpha_1, \alpha_2\}$.

Let us compare the functions included in Eq. (2.4) with their images:

$$K(\tau) \doteqdot \widetilde{K}(\rho), \quad f(x) \doteqdot F(p), \quad \varphi(x) \doteqdot \Phi(p),$$

where

$$F(p) = \int_0^\infty f(x)e^{-px}\mathrm{d}x,$$

$$\widetilde{K}(p) = \int_0^\infty K(\tau)e^{-p\tau}d\tau$$

are known functions of the complex variable p, and the image $\Phi(p)$ is unknown since the function $\varphi(x)$ is unknown.

By the convolution theorem, $\widetilde{K}(p) \cdot \Phi(p)$ is an image of the convolution of functions $(\varphi * K)$, i.e.,

$$\widetilde{K}(p) \cdot \Phi(p) \doteqdot (\varphi * K) = \int_0^x K(x - t)\varphi(t)\mathrm{d}t.$$

Applying the Laplace transform to both sides of the Volterra equation of the second kind with a difference kernel, i.e., passing in the integral Eq. (2.4) to images and using the convolution theorem, we obtain the algebraic equation

$$\Phi(p) = \lambda \widetilde{K}(p) \cdot \Phi(\rho) + F(p),$$

from where we can express the desired image

$$\Phi(p) = \frac{F(p)}{1 - \lambda \widetilde{K}(\rho)}.$$

Since $\Phi(p)$ is an analytic function in the half-plane $\mathrm{Re}(p) = s > \max\{\alpha_1, \alpha_2\}$, the denominator of the fraction in the last equality cannot be equal to zero in this half-plane. The original of the image $\Phi(p)$ is a solution of the original integral equation. The solution is unique for any appropriate λ. The original $\varphi(x)$ can be expressed by the Riemann–Mellin inversion formula:

$$\varphi(x) = \frac{1}{2\pi i} \int\limits_{s-i\infty}^{s+i\infty} \frac{F(p)}{1 - \lambda \widetilde{K}(p)} e^{pt} \mathrm{d}p, \quad s > \alpha_3.$$

However, in many cases the original is easier to find by using properties and tables of originals and images, or using symbolic computation packages such as *Wolfram Mathematica*.

Example 2.3 Solve the equation

$$\varphi(x) = x + \int\limits_0^x \sin(x - t)\varphi(t)\mathrm{d}t.$$

Solution
Let us match

$$\varphi(x) \doteqdot \Phi(p), \quad f(x) = x \doteqdot \frac{1}{p^2}, \quad K(\tau) = \sin\tau \doteqdot \frac{1}{p^2 + 1}.$$

Then, the equation in the images takes the following form

$$\Phi(p) = \frac{1}{p^2} + \frac{1}{p^2 + 1}\Phi(p),$$

where

$$\Phi(p) = \frac{p^2 + 1}{p^4} = \frac{1}{p^2} + \frac{1}{p^4}.$$

From the table, we find the original which is the only solution to the equation:

$$\varphi(x) = x + \frac{x^3}{3!} = x + \frac{x^3}{6}.$$

The same steps can be done in *Wolfram Mathematica*. First, set the initial functions and the integral equation.

```
In [1]:= Clear[f,K,φ,F,KK,Φ,x,p]
         f[x_]:=x
         K[τ_]:= Sin[τ]
         inteqn=φ[x]==f[x]+Integrate[K[x-t]φ[t],{t,0,x}]
Out [4]:= φ[x]==x+ ∫₀ˣ −Sin[t−x]φ[t]dt
```

Then, find images for known functions and compose an algebraic equation.

```
In [5]:= F[p_]:= LaplaceTransform[f[x],x,p]
         KK[p_]:=LaplaceTransform[K[τ],τ,p]
         imageeqn=Φ[p]==F[p]+KK[p]Φ[p]
Out [7]:=Φ[p]== 1/p² + Φ[p]/(1+p²)
```

After that, it remains to solve the resulting equation and restore the original solution.

```
In [8]:= imagesol=Solve[imageeqn,Φ[p]][[1,1]]
Out [8]:=Φ[p]→ (1+p²)/p⁴
In [9]:= sol=φ->Function[{x},
         Evaluate[InverseLaplaceTrans form
         [Φ[p]/.imagesol,p,x]]]
Out [9]:=φ→ Function[{x},x+ x³/6 ]
```

Let us check that this function is the solution.

```
In [10]:= inteqn/.sol
Out [10]:=True
```

Answer

$$\varphi(x) = x + \frac{x^3}{6}$$

Example 2.4 Solve integral equation

$$\varphi(x) = \sin x + \lambda \int_0^x \cos(x - t)\varphi(t)dt$$

when the parameter $\lambda = 2$.

Solution
A solution of the equation will be found in *Wolfram Mathematica*. Set the functions and the original equation.

```
In [11]:= Clear[f,K,φ,λ,Φ,x,p]
          f[x_]:=Sin[x]
          K[τ_]:=Cos[τ]
          λ = 2;
          inteqn=φ[x]==f[x]+λ Integrate[K[x-t]φ[t],{t,0,x}]
Out [15]:=φ[x]== 2∫₀ˣ Cos[t−x]φ[t]dt+Sin[x]
```

Then, apply the Laplace transform to the original integral equation and construct an algebraic equation.

```
In [16]:= imageeqn=LaplaceTransform[inteqn,x,p]/.
          LaplaceTransform[φ[x],x,p]->Φ[p]
Out [16]:=Φ[p]== 1/(1+p²) + 2pΦ[p]/(1+p²)
```

And finally, solve the resulting algebraic equation and restore the original.

```
In [17]:= imagesol=Solve[imageeqn,Φ[p]][[1,1]]
Out [17]:=Φ[p]→ 1/(−1+p)²
In[18]:= sol=φ->Function[{x},
```

```
          Evaluate[InverseLaplaceTransform[Φ[p]/.imagesol,p,x]]]
```
Out[18]:=$\phi\rightarrow$ **Function**$[\{x\}, e^x x]$

Let us check the solution by substituting it into the original integral equation.

```
   In [19]:= inteqn/.sol
```
Out[19]:=**True**

You can write your own function that allows to solve integral equations of this type with an arbitrary difference kernel and a free term.

```
In [20]:= Clear[ISolveLaplace]
          ISolveLaplace[K_,f_,φ_,λ_,x_]:=
          Block[{inteqn,imageeqn,imagesol,Φ,p},
          inteqn=φ[x]==f[x]+λ Integrate[K[x-t]φ[t],{t,0,x}];
          imageeqn=LaplaceTransform[inteqn,x,p]/.
          LaplaceTransform[φ[x],x,p]->Φ[p];
          imagesol=Solve[imageeqn,Φ[p]][[1,1]];
          φ->Function[{t},Evaluate[InverseLaplaceTransform[Φ[p]/.
          imagesol,p,t]]]/.t->x
          ]
```

Calling this function and substituting the free term and the kernel of the integral equation as pure functions, we easily obtain the solution.

```
   In [22]:= Clear[f,K,φ,λ,x]
             f=Function[{x},Sin[x]];
             K=Function[{τ},Cos[τ]];
             λ=2;
             sol=ISolveLaplace[K,f,φ,λ,x]
```
Out[26]:=$\phi\rightarrow$ **Function**$[\{x\}, e^x x]$

Let us plot the graph of the solution (Fig.2.2)

```
   In [27]:=optplot={AxesStyle->Arrowheads[{0.0,0.025}],
            AxesLabel->{Style[x,Bold,Medium],Style[y,Bold,Medium]}};
            Plot[φ[x]/.sol,{x,-5,5},Evaluate@optplot]
```

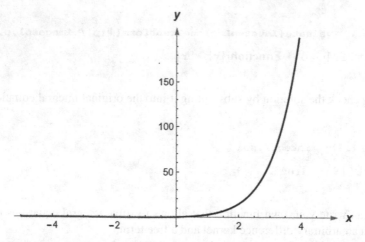

Fig. 2.2 Graph of the solution of an integral equation with a difference kernel (Example 2.4)

Note. For this equation, we can consider the general case for an arbitrary value of the parameter λ:

```
In [29]:= Clear[f,K,φ,λ,x]
          f=Function[{x},Sin[x]];
          K=Function[{τ},Cos[τ]];
          sol=ISolveLaplace[K,f,φ,λ,x]

Out[32]:=φ→ Function[ {x}, −
          e^{x(\frac{λ}{2}-\frac{1}{2}\sqrt{-4+λ^2})} − e^{x(\frac{λ}{2}+\frac{1}{2}\sqrt{-4+λ^2})}
          ───────────────────────────────────────────────
                          \sqrt{-4+λ^2}                    ]
```

Let us plot the graph of the solution (Fig. 2.2).

Note that the case $\lambda = \pm 2$ turns out to be special, but it can be obtained from the general solution by computing the limit $\lambda \to \pm 2$.

Answer

$$\varphi(x) = e^x x, \quad \text{if} \quad \lambda = 2$$

$$\varphi(x) = -\frac{e^{x\left(\frac{\lambda}{2}-\frac{1}{2}\sqrt{-4+\lambda^2}\right)} - e^{x\left(\frac{\lambda}{2}+\frac{1}{2}\sqrt{-4+\lambda^2}\right)}}{\sqrt{-4+\lambda^2}}, \quad \text{if} \quad \lambda \neq \pm 2$$

Example 2.5 Solve integral equation

$$\varphi(x) = x - \lambda \int\limits_0^x (x - t)\varphi(t)dt.$$

Solution
We have already solved this equation in the Example 2.1 using the built-in function **DSolve**. This time, we will use the **ISolveLaplace** function that we have written.

```
In [33]:= Clear[f,K,φ,λ,x]
          f=Function[{x},x];
          K=Function[{τ},-τ];
          sol=ISolveLaplace[K,f,φ,λ,x]]
```

$$\text{Out [36]} := \phi \rightarrow \textbf{Function}\left[\{x\}, \frac{\text{Sin}\left[x\sqrt{\lambda}\right]}{\sqrt{\lambda}}\right]$$

The solution using the Laplace transform coincides with the solution using the **DSolve** function.

Answer
The same answer as in Example 2.1b.

$$\varphi(x) = \frac{\text{Sin}\left(x\sqrt{\lambda}\right)}{\sqrt{\lambda}}$$

Example 2.6 Solve integral equation

$$\varphi(x) = \cos x + \lambda \int\limits_0^x \sin(x - t)\varphi(t)dt.$$

Solution
We set the free term and the kernel and call the **ISolveLaplace** function.

```
In [37]:= Clear[f,K,φ,λ,x]
          f=Function[{x},Cos[x]];
          K=Function[{τ},Sin[τ]];
          sol=ISolveLaplace[K,f,φ,λ,x]]
```

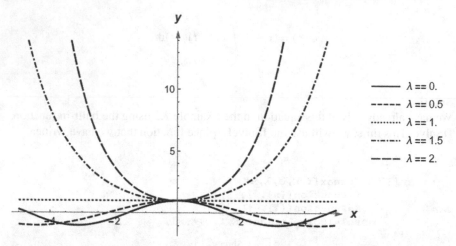

Fig. 2.3 Family of solutions of an integral equation with a difference kernel for different values of the parameter λ (Example 2.6)

Out [40] :=$\phi\rightarrow$ **Function**$\left[\{x\},\,\text{Cos}\left[x\sqrt{1-\lambda}\,\right]\right]$

Let us plot the graph of the found solution for different values of (Fig. 2.3)λ.

```
In [41]:= Plot[Table[φ[x]/.sol,{λ,0,2,0.5}]//Evaluate,{x,-5,5}
           ,PlotLegends->ToString/@Thread[λ==Range[0,2,0.5]]]
```

Answer

$$\varphi(x) = \text{Cos}\left(x\sqrt{1-\lambda}\right)$$

2.2.2 Solving Systems of Volterra Integral Equations with Difference Kernels Using the Laplace Transform

The Laplace transform can be used to solve systems of integral equations with difference kernels as

$$\varphi_i(x) = f_i(x) + \sum_{j=1}^{s} \int_0^x K_{ij}(x-t)\varphi_j(t)dt, \quad i = 1, 2, \ldots, s, \qquad (2.5)$$

where $K_{ij}(x)$, $f_i(x)$ are known continuous functions having a Laplace image.

Applying the Laplace transform to system (2.5), we obtain

$$\Phi_i(p) = F_i(p) + \sum_{j=1}^{s} \widetilde{K}_{ij}(p)\Phi_j(p), \quad i = 1, 2, \ldots, s. \qquad (2.6)$$

This is a system of linear algebraic equations for $\Phi_i(p)$. Having solved it, we find $\Phi_i(p)$, the originals for which will be the solution of the original system of integral Eqs. (2.5).

Example 2.7 Solve the system of integral equations

$$\begin{cases} \varphi_1(x) = 1 - 2\int_0^x e^{2(x-t)}\varphi_1(t)dt + \int_0^x \varphi_2(t)dt \\ \varphi_2(x) = 4x - \int_0^x \varphi_1(t)dt + \int_0^x (x-t)\varphi_2(t)dt \end{cases}.$$

Solution

If we apply the Laplace transform to this system, we get the algebraic system

$$\begin{cases} \Phi_1(p) = F_1(p) + \widetilde{K}_{11}(p)\Phi_1(p) + \widetilde{K}_{12}(p)\Phi_2(p) \\ \Phi_2(p) = F_2(p) + \widetilde{K}_{21}(p)\Phi_1(p) + \widetilde{K}_{22}(p)\Phi_2(p) \end{cases}.$$

This system could be solved manually, but we will solve it in *Wolfram Mathematica*. Let us set the vector of free terms and the matrix of kernels.

```
In [1]:=Clear[fList,KList,φList,F,KK,Φ,x,p]
        fList[x_]={1,4x};
        KList[τ_]={{-2Exp[2τ],1},{-1,4τ}};
        φList[x_]={φ1[x],φ2[x]};
        intsyseqn=Thread[φList[x]==fList[x]+
        Integrate[KList[x-t].φList[t],{t,0,x}]]/.
        {Integrate[f_+g_,t_]:>Integrate[f,t]+Integrate[g,t]}
```

$$\text{Out [5] :=}\{\phi1[x]== 1 + \int_0^x -2e^{2(-t+x)}\phi1[t]dt + \int_0^x \phi2[t]dt,$$

$$\phi2[x]== 4x + \int_0^x -\phi1[t]dt + \int_0^x 4(-t+x)\phi2[t]dt\}$$

It should be noted that the built-in function **DSolve** cannot solve systems of integral equations.

```
In [6] := DSolve[intsyseqn,{φ1,φ2},x][[1]]
```
Out[6] :=

$$\left\{ \phi1 \rightarrow \mathbf{Function}\left[\{x\}, 1 + \int_0^x -2e^{2(-t+x)}\phi1[t]dt + \int_0^x \phi2[t]dt \right], \right.$$

$$\left. \phi2 \rightarrow \mathbf{Function}\left[\{x\}, 4x + \int_0^x -\phi1[t]dt + \int_0^x 4(-t+x)\phi2[t]dt \right] \right\}$$

So, we obtain images and solve the algebraic system.

```
In [7] := ΦList[p_]={Φ1[x],Φ2[x]};
          imagesyseqn=LaplaceTransform[intsyseqn,x,p]/.
          Thread[LaplaceTransform[φList[x],x,p]->ΦList[p]]
```
Out[8] := $\left\{ \Phi1[x] == \frac{1}{p} - \frac{2\Phi1[x]}{-2+p} + \frac{\Phi2[x]}{p}, \; \Phi2[x] == \frac{4}{p^2} - \frac{\Phi1[x]}{p} + \frac{4\Phi2[x]}{p^2} \right\}$

```
In [9] := imagesyssol=Solve[imagesyseqn,ΦList[p]][[1]]
```
Out[9] := $\left\{ \Phi1[x] \rightarrow \frac{p}{(1+p)^2}, \; \Phi2[x] \rightarrow -\frac{-2-3p}{(-2+p)(1+p)^2} \right\}$

Then, we find the originals for the obtained images and check that they are the solutions.

```
In [10] := syssol=Thread[{φ1,φ2}->(Function[{x},#]&/@
           InverseLaplaceTransform[ΦList[p]/.imagesyssol,p,x])]
```
Out[10] := $\{\phi1 \rightarrow \mathbf{Function}[\{x\}, -e^{-x}(-1+x)],$

$$\phi2 \rightarrow \mathbf{Function}\left[\{x\}, \frac{1}{9}e^{-x}(-8 + 8e^{3x} + 3x) \right] \}$$

```
In [11] := intsyseqn/.syssol//FullSimplify
```
Out[11] := {**True, True**}.

All described steps can be implemented as a function for systems of arbitrary dimensions.

```
In [12]:= Clear[ISolveSysLaplace]
          ISolveSysLaplace[n_,fList_,KList_,x_]:=
          Block[{φList,ΦList,FList,KKList,imagesyseqn,
          imagesyssol,τ,p,φ,Φ,t},
          φList=Function[{t},Subscript[φ,#][t]&/@Range[n]];
          ΦList=Function[{t},Subscript[Φ,#][t]&/@Range[n]];
          FList=LaplaceTransform[fList[τ],τ,p];
          KKList=LaplaceTransform[KList[τ],τ,p];
      imagesyseqn=Thread[ΦList[p]==FList+KKList.ΦList[p]];
          imagesyssol=Solve[imagesyseqn,ΦList[p]][[1]];
          Thread[Subscript[φ,#]&/@Range[n]->
      (Function[{t},#]&/@InverseLaplaceTransform[ΦList[p]/.
          imagesyssol,p,t])]/.t->x
          ]
```

Let us check the defined function. Now, the vector of free terms and the matrix of the kernels of the equations are specified in the form of pure functions.

```
In [14]:= Clear[fList,KList,x]
          fList=Function[{x},{1,4x}];
          KList=Function[{τ},{{-2Exp[2τ],1},{-1,4τ}}];
          sol=ISolveSysLaplace[2,fList,KList,x]
```
$$Out[17]:-\{\phi1 \rightarrow \text{Function}\big[\{x\},-e^{-x}(-1+x)\big],$$
$$\phi2 \rightarrow \text{Function}\Big[\{x\},\frac{1}{9}e^{-x}(-8+8e^{3x}+3x)\Big]\}$$

The answer is the same as for the step-by-step solution.
Let us plot the graphs of the solutions (Fig. 2.4).

```
In [18]:= φList=Function[{t},Subscript[φ,#][t]&/@Range[2]]
          Plot[φList[x]/.sol//Evaluate,{x,-5,5},
          PlotLegends->φList[x]]
```

Answer

$$\begin{cases} \varphi_1(x) = -e^{-x}(-1+x) \\ \varphi_2(x) = \frac{1}{9}e^{-x}(-8+8e^{3x}+3x) \end{cases}$$

Example 2.8 Solve the system of integral equations

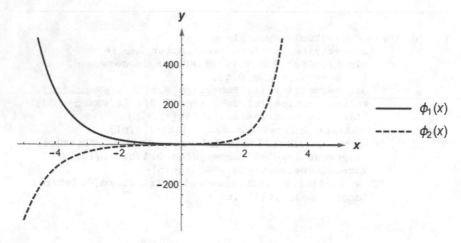

Fig. 2.4 Solution for a system of integral equations with a difference kernel (Example 2.7)

$$\begin{cases} \varphi_1(x) = 1 - \int\limits_0^x \varphi_2(t)\mathrm{d}t \\[2mm] \varphi_2(x) = \cos x - 1 + \int\limits_0^x \varphi_3(t)\mathrm{d}t \ . \\[2mm] \varphi_3(x) = \cos x + \int\limits_0^x \varphi_1(t)\mathrm{d}t \end{cases}$$

Solution

Let us use the **ISolveSysLaplace** function written in the previous example. First, we set the vector of free terms and the matrix of the kernels of the equations in the form of pure functions. And then, we call the function **ISolveSysLaplace**.

```
In [20]:= Clear[n,fList,KList,x]
          n=3;
          fList=Function[{x},{1,Cos[x]-1,Cos[x]}];
          KList=Function[{τ},{{0,-1,0},{0,0,1},{1,0,0}}];
          sol=ISolveSysLaplace[n,fList,KList,x]

Out[24]:={ϕ₁→ Function[{x}, Cos[x]],
           ϕ₂→ Function[{x}, Sin[x]]
           ϕ₃→ Function[{x}, Cos[x]+Sin[x]]
```

Let us check the solutions by substituting them into the original system.

```
In [25]:= φList=Function[{t},Subscript[φ,#][t]&/@Range[n]];
         Thread[φList[x]==fList[x]+
         Integrate[KList[x-t].φList[t],{t,0,x}]]/.sol

Out[26]:={True, True, True}
```

Answer

$$\begin{cases} \varphi_1(x) = \cos x \\ \varphi_2(x) = \sin x \\ \varphi_3(x) = \cos x + \sin x \end{cases}$$

2.2.3 Solving Integro-Differential Equations with Difference Kernels Using the Laplace Transform

An equation

$$a_0(x)\varphi^{(n)}(x) + a_1(x)\varphi^{(n-1)}(x) + \cdots + a_n(x)\varphi(x)$$
$$+ \sum_{m=0}^{s} \int_0^x K_m(x,t)\varphi^{(m)}(t)dt = f(x), \tag{2.7}$$

in which the coefficients $a_0(x),\ldots, a_n(x)$, the free term $f(x)$ and the kernels $K_m(x,t)$ are known functions, and $\varphi(x)$ is the required one, is called a *linear integro-differential equation* [2].

Solving the integro-differential Eq. (2.7), unlike an integral one, for the required function, it is necessary to set the initial conditions:

$$\varphi(0) = \varphi_0, \quad \varphi'(0) = \varphi'_0, \quad \ldots, \quad \varphi^{(n-1)}(0) = \varphi_0^{(n-1)}. \tag{2.8}$$

Consider the integro-differential Eq. (2.7) with constant coefficients and difference kernels. We assume that all coefficients $a_i(x) = a_i$ and the kernels have the form $K_m(x,t) = K_m(x-t)$. Without loss of generality, we can assume that $a_0(x) = 1$, then Eq. (2.7) turns to

$$\varphi^{(n)}(x) + a_1\varphi^{(n-1)}(x) + \cdots + a_n\varphi(x) + \sum_{m=0}^{s} \int_0^x K_m(x-t)\varphi^{(m)}(t)dt = f(x)$$

$$\tag{2.9}$$

Perform the Laplace transform of this equation. Let the functions $f(x)$ and $K_m(x)$ be original functions, then their images are

$$f(x) \doteqdot F(p), \quad K(x) \doteqdot \widetilde{K}(p).$$

In this case, the unknown function $\varphi(x)$ also has a Laplace image

$$\varphi(x) \doteqdot \Phi(p).$$

We apply the Laplace transform to both sides of the equation. The theorem about the image of the derivative says that the transformation for the derivatives can be written as:

$$\varphi^{(k)}(x) \doteqdot p^k \Phi(p) - p^{k-1}\varphi_0 - \ldots - \varphi_0^{(k-1)}.$$

The multiplication theorem gives:

$$\int_0^x K(x-t)\varphi^{(m)}(t)dt \doteqdot \widetilde{K}_m(p)\left[p^m \Phi(p) - p^{m-1}\varphi_0 - \ldots - \varphi_0^{(m-1)} \right].$$

Thus, the original equation is converted to

$$\Phi(p)\left[p^n + a_1 p^{n-1} + \ldots + a_n + \sum_{m=0}^{s} \widetilde{K}_m(p)p^m \right] = A(p),$$

where $A(p)$ is some known function of p.

From the last algebraic equation, we find the image Φ (p), find the original for it and obtain $\varphi(x)$ which is the solution of the integro-differential Eq. (2.9) with the given initial conditions (2.8).

Some of the simplest cases of integro-differential equations can be solved using the *Wolfram Mathematica*.

Example 2.9 Solve integro-differential equation with parameter

$$\varphi'(x) = 1 + \sin(cx) + \int_0^x \varphi(t)dt,$$

if $\varphi(0) = -1$.

Solution

Let us try to find a solution in Wolfram Mathematica using the built-in **DSolve** function. Set all the original functions, as well as the initial condition in the form of an equation.

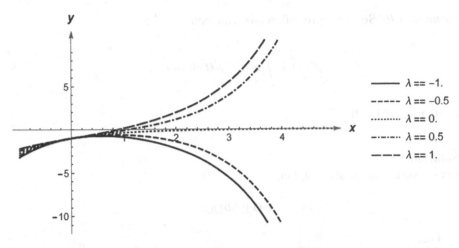

Fig. 2.5 Family of solutions of an integro-differential equation with a parameter (Example 2.9)

```
In [1]:= Clear[f,K]
        f[x_]=1+Sin[c x];
     K[x_,t_]=1;
        initcond=φ[0]==-1;
        eqn=φ'[x]==f[x]+Integrate[K[x,t]φ[t],{t,0,x}]
```

$$\text{Out}[5]:=\phi'[x]== 1+\int_0^x \phi[t]dt+\text{Sin}[cx]$$

Find a solution. It contains the parameter c, so we will construct the plots of the family of solutions for different values of the parameter c (Fig. 2.5).

```
In [6]:= DSolve[{eqn,initcond},φ,x][[1,1]]
```

$$\text{Out}[6]:=\phi\rightarrow \textbf{Function}\left[\{x\}, \frac{e^{-x}(-2+c-2c^2+ce^{2x}-2ce^x\text{Cos}[cx])}{2(1+c^2)}\right]$$

```
In [7]:= optplot={AxesStyle->Arrowheads[{0.0,0.025}],
        AxesLabel->{Style[x,Bold,Medium],Style[y,Bold,Medium]}}
        ;Plot[Table[φ[x]/.sol,{c,-1,1,0.5}]//Evaluate,{x,-1,5},
        PlotLegends->ToString/@Thread[λ==Range[-1,1,0.5]],
        Evaluate@optplot]
```

Answer

$$\varphi(x) = \frac{e^{-x}(-2+c-2c^2+ce^{2x}-2ce^x\text{Cos}(cx))}{2(1+c^2)}$$

Example 2.10 Solve integro-differential equation

$$\varphi''(x) + \int\limits_0^x e^{2(x-t)}\varphi'(t)\mathrm{d}t = e^{2x},$$

if $\varphi(0) = \varphi'(0) = 0$.

Solution

Let us solve analytically first. Let

$$\varphi(x) \fallingdotseq \Phi(p).$$

Then,

$$\varphi'(x) \fallingdotseq p\Phi(p)$$

and

$$\varphi''(x) \fallingdotseq p^2\Phi(p).$$

After applying the Laplace transform for the equation, we get

$$p^2\Phi(p) + \frac{p}{p-2}\Phi(p) = \frac{1}{p-2}.$$

For the image of unknown function,

$$\Phi(p) = \frac{1}{p(p-1)^2} = \frac{1}{p} + \frac{2}{(p-1)^2} - \frac{p}{(p-1)^2}$$

and for the original:

$$\varphi(x) = xe^x - e^x + 1.$$

Thus, a solution has been found.

Now, let us implement the same solution in *Wolfram Mathematica*.

Set the initial data in the form of a list of coefficients at derivatives, a list of integral kernels, a free term and initial conditions.

```
In [9]:= Clear[f,KList]
         f=Function[{x},Exp[2 x]];
         KList=Function[{τ},{0,Exp[2τ],0}];
```

```
aList={0,0,1};
initcond={φ[0]==0,φ'[0]==0};
dφList=Function[{x},D[φ[x],{x,#}]&/@
Range[0,Length@aList-1]];
eqn=dφList[x].aList+
Integrate[KList[x-t].dφList[t],{t,0,x}]==f[x]
```

$$\text{Out}[15]:=\int_0^x e^{2(-t+x)}\phi'[t]dt+\phi''[x]==e^{2x}$$

Is easy to see that the built-in **DSolve** function does not handle this equation.

```
In [16]:= DSolve[{eqn,initcond},φ,x][[1,1]]
```

$$\text{Out}[16]:=\int_0^x e^{2(-t+x)}\phi'[t]dt+\phi''[x]==e^{2x}$$

Therefore, we apply the Laplace transform. We transform the initial conditions into transformation rules so that it is convenient to substitute them, and we obtain an algebraic equation for the desired image.

```
In [17]:= initcondrule=initcond/.Equal->Rule
```

$$\text{Out}[17]:=\left\{\phi[0]\to 0,\phi'[0]\to 0\right\}$$

```
In [18]:= imageeqn=LaplaceTransform[eqn,x,p]/.
        LaplaceTransform[φ[x],x,p]->Φ[p]/.initcondrule
```

$$\text{Out}[18]:=\frac{p\Phi[p]}{-2+p}+p^2\Phi[p]==\frac{1}{-2+p}$$

Solve the resulting equation and restore the original.

```
In [19]:= imagesol=Solve[imageeqn,Φ[p]][[1,1]]
```

$$\text{Out}[19]:=\Phi[p]\to\frac{1}{(-1+p)^2 p}$$

```
In [20]:= sol=φ->Function[{x},InverseLaplaceTransform[Φ[p]/.
        imagesol,p,x]//Evaluate]
```

$$\text{Out}[20]:=\phi\to \text{Function}[\{x\},1+e^x(-1+x)]$$

The result is the same as the manual solution.
Let us check the solution by substitution to the equation.

```
In [21]:= eqn/.sol//FullSimplify

Out[21]:=True
```

These steps can be implemented as a function.

```
In [22]:= Clear[IDSolveLaplace]
          IDSolveLaplace[f_,KList_,aList_,initcond_,φ_,x_]:=
          Block[{dφList,eqn,initcondrule
          ,imageeqn,imagesol,p,t,Φ},
          dφList=Function[{t},D[φ[t],{t,#}]]&/@
          Range[0,Length@aList-1]];
          eqn=dφList[x].aList+
          Integrate[KList[x-t].dφList[t],{t,0,x}]==f[x]/.
          {Integrate[f1_+f2_,var_]:>
          Integrate[f1,var]+Integrate[f2,var]};
          initcondrule=initcond/.Equal->Rule;
          imageeqn=LaplaceTransform[eqn,x,p]/.
          LaplaceTransform[φ[x],x,p]->Φ[p]/.initcondrule;
          imagesol=Solve[imageeqn,Φ[p]][[1,1]];
          φ->Function[{t},InverseLaplaceTransform[Φ[p]/.
          imagesol,p,t]//Evaluate]/.t->x
          ]
```

Let us check this function on the already specified initial data.

```
In [24]:= Clear[φ,x]
          IDSolveLaplace[f,KList,aList,initcond,φ,x]

Out[25]:=φ→ Function[{x}, 1+e^x(-1+x)]
```

Answer

$$\varphi(x) = xe^x - e^x + 1$$

Example 2.11 Solve integro-differential equation

$$\varphi''(x) - 2\varphi'(x) + \varphi(x) + 2 \int\limits_0^x \cos(x - t)\varphi''(t)\mathrm{d}t$$

$$+2\int_0^x \sin(x - t)\varphi'(t)dt = \cos(x)$$

under the initial conditions $\varphi(0) = a, \varphi'(0) = b$.

Solution
Set the initial data and apply the previously written **IDSolveLaplace** function.

```
In [26]:= Clear[f,KList,aList,initcond,φ,x,a,b]
          f=Function[{x},Cos[x]];
          KList=Function[{τ},{0,Sin[τ],Cos[τ]}];
          aList={1,-2,1};
          initcond={φ[0]==a,φ'[0]==b};
          dφList=Function[{x},D[φ[x],{x,#}]&/@
          Range[0,Length@aList-1]];
          eqn=dφList[x].aList+
          Integrate[KList[x-t].dφList[t],{t,0,x}]==f[x]/.
          {Integrate[f1_+f2_,var_]:>
          Integrate[f1,var]+Integrate[f2,var]}
```

$$\text{Out}[32]:= \int_0^x -\text{Sin}[t-x]\phi'[t]dt$$

$$+ \int_0^x \text{Cos}[t-x]\phi''[t]dt+\phi[x]-2\phi'[x]+\phi''[x]$$

$$== \text{Cos}[x]$$

```
In [33]:= sol=IDSolveLaplace[f,KList,aList,initcond,φ,x]
```

$$\text{Out}[33]:=\phi\rightarrow \text{Function}\Big[\{x\},(-1-b)\text{Sin}[x]$$

$$+\frac{e^{x/2}\left(\sqrt{3}a\text{Cos}\left[\frac{\sqrt{3}x}{2}\right]+2\text{Sin}\left[\frac{\sqrt{3}x}{2}\right]-a\text{Sin}\left[\frac{\sqrt{3}x}{2}\right]+4b\text{Sin}\left[\frac{\sqrt{3}x}{2}\right]\right)}{\sqrt{3}}\Big]$$

Check the obtained solution by substitution and construct plots of the family of curves for various initial conditions (Fig. 2.6).

```
In [34]:= eqn/.sol//FullSimplify
```

Out[34]:=**True**

```
In [35]:=Plot[Table[φ[x]/.sol,{a,-1,1},{b,0,1}]//
         Evaluate,{x,-1,8},PlotRange->All,PlotLegends->
```

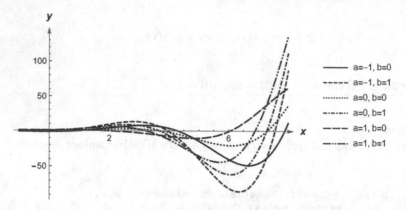

Fig. 2.6 Family of solutions of an integro-differential equation with different initial conditions (Example 2.11)

```
Flatten@Table["a="<>ToString[a]<>", b="<>ToString[b],
    {a,-1,1},{b,0,1}],Evaluate@optplot]
```

Answer

$$\varphi(x) = (-1 - b)\mathrm{Sin}\,x +$$

$$\frac{e^{x/2}\left(\sqrt{3}a\cos\left(\frac{\sqrt{3}x}{2}\right) + 2\sin\left(\frac{\sqrt{3}x}{2}\right) - a\sin\left(\frac{\sqrt{3}x}{2}\right) + 4b\sin\left(\frac{\sqrt{3}x}{2}\right)\right)}{\sqrt{3}}$$

2.3 Solving Fredholm Integral Equations with Difference Kernels Using the Fourier Transform

Consider a Fredholm equation of the second kind with a difference kernel and infinite limits of integration

$$\varphi(x) = f(x) + \lambda \int_{-\infty}^{+\infty} K(x - t)\varphi(t)\mathrm{d}t. \tag{2.10}$$

Let the function $f(x)$ be square integrable, and the function $K(\tau)$ be absolutely integrable on the whole axis, i.e., integrals exist

$$\int\limits_{-\infty}^{+\infty} f^2(x)\mathrm{d}x < +\infty$$

and

$$\int\limits_{-\infty}^{+\infty} |K(\tau)|\mathrm{d}\tau < +\infty.$$

To solve this equation, the Fourier transform is usually used [7–9]. Let us find the Fourier transforms of the functions $f(x)$, $\varphi(x)$, $K(\tau)$:

$$F(\omega) = \frac{1}{\sqrt{2\pi}} \int\limits_{-\infty}^{+\infty} f(x)e^{-i\omega x}\mathrm{d}x;$$

$$\Phi(\omega) = \frac{1}{\sqrt{2\pi}} \int\limits_{-\infty}^{+\infty} \varphi(x)e^{-i\omega x}\mathrm{d}x;$$

$$\tilde{K}(\omega) = \frac{1}{\sqrt{2\pi}} \int\limits_{-\infty}^{+\infty} K(\tau)e^{-i\omega\tau}\mathrm{d}t,$$

where ω is a real parameter of transformations. By definition, the integral

$$(\varphi * K) = \int\limits_{-\infty}^{+\infty} K(x - t)\varphi(t)\mathrm{d}t$$

we call a convolution of functions. By the convolution theorem for the Fourier transform, we have

$$(\varphi * K) \dot= \sqrt{2\pi}\,\tilde{K}(\omega)\Phi(\omega).$$

Passing in Eq. (2.10) to the Fourier images, we obtain

$$\Phi(\omega) = F(\omega) + \lambda\sqrt{2\pi}\,\tilde{K}(\omega)\Phi(\omega)$$

where under the condition

$$1 - \lambda\sqrt{2\pi}\,\tilde{K}(\omega) \neq 0 \qquad\qquad (2.11)$$

we get

$$\Phi(\omega) = \frac{F(\omega)}{1 - \lambda\sqrt{2\pi}\,\widetilde{K}(\omega)}.$$

The solution of Eq. (2.10) is obtained using the inverse Fourier transform

$$\varphi(x) = \frac{1}{\sqrt{2\pi}} \int\limits_{-\infty}^{+\infty} \frac{F(\omega)}{1 - \lambda\sqrt{2\pi}\,\widetilde{K}(\omega)} e^{i\omega x} d\omega. \tag{2.12}$$

It is unique under condition (2.11), and $\varphi(x)$ has a finite norm in L_2, so

$$\int\limits_{-\infty}^{+\infty} \varphi^2(x)dx < +\infty.$$

The Fourier transform can also be used to solve the Fredholm integral equations of the first kind with a difference kernel and infinite limits of integration

$$\int\limits_{-\infty}^{+\infty} K(x - t)\varphi(t)dt = f(x).$$

If we use the convolution theorem, we can write the equation in the Fourier images

$$\sqrt{2\pi}\,\widetilde{K}(\omega)\Phi(\omega) = F(\omega),$$

where

$$\Phi(\omega) = \frac{1}{\sqrt{2\pi}} \frac{F(\omega)}{\widetilde{K}(\omega)}.$$

The unique solution of the equation is written as

$$\varphi(x) = \frac{1}{2\pi} \int\limits_{-\infty}^{+\infty} \frac{F(\omega)}{\widetilde{K}(\omega)} e^{i\omega x} d\omega,$$

if $f(x)$ has a finite norm, $K(\tau)$ is an absolutely integrable function on the whole axis, and the condition

$$\int\limits_{-\infty}^{+\infty} \left[\frac{F(\omega)}{\widetilde{K}(\omega)}\right]^2 d\omega < +\infty$$

is satisfied.

Along with the Fourier transform, sine and cosine Fourier transforms are often used to solve Fredholm integral equations with a difference kernel over a semi-infinite integration interval. The sine Fourier transform of a function $\varphi(t)$ is a function of a real variable x defined as

$$\Phi_s(x) = \sqrt{\frac{2}{\pi}} \int_0^\infty \varphi(t) \sin xt dt.$$

Similarly, the cosine transform of functions is the function of the form

$$\Phi_c(x) = \sqrt{\frac{2}{\pi}} \int_0^\infty \varphi(t) \cos xt dt.$$

Inverse sine and cosine Fourier transforms are calculated by the formulas:

$$\varphi(t) = \sqrt{\frac{2}{\pi}} \int_0^\infty \Phi_s(x) \sin xt \, dx; \quad \varphi(t) = \sqrt{\frac{2}{\pi}} \int_0^\infty \Phi_c(x) \cos xt \, dx.$$

Example 2.12 Solve the Fredholm equation of the second kind

$$\varphi(x) = e^{-|x|} + \lambda \int_{-\infty}^{+\infty} K(x - t)\varphi(t)dt,$$

where

$$K(\tau) = \begin{cases} e^\tau, & \tau < 0 \\ 0, & \tau > 0 \end{cases}.$$

Solution
Find the image $f(x)$ and $K(\tau)$ by the Fourier transform

$$F(\omega) = \frac{1}{\sqrt{2\pi}} \int_{-\infty}^{+\infty} e^{-|x|-i\omega x} dx = \frac{1}{\sqrt{2\pi}} \left[\int_{-\infty}^0 e^{x(1-i\omega)} dx + \int_0^{+\infty} e^{-x(1+i\omega)} dx \right]$$

$$= \frac{1}{\sqrt{2\pi}} \left[\frac{1}{1-i\omega} e^{x(1-i\omega)} \Big|_{-\infty}^0 - \frac{1}{1+i\omega} e^{-x(1+i\omega)} \Big|_0^{+\infty} \right] = \sqrt{\frac{2}{\pi}} \frac{1}{1+\omega^2}$$

where it was taken into account that

$$\left| e^{\pm x(1-i\omega)} \right| = e^{\pm x} \to 0, \qquad x \to \mp\infty.$$

and

$$\tilde{K}(\omega) = \frac{1}{\sqrt{2\pi}} \int\limits_{-\infty}^{+\infty} K(\tau)e^{-i\omega\tau}\,d\tau = \frac{1}{\sqrt{2\pi}} \int\limits_{-\infty}^{0} e^{\tau(t-i\omega)}\,d\tau = \frac{1}{\sqrt{2\pi}}\frac{1}{1-i\omega}$$

The solution of the equation by formula (2.12) has the form

$$\varphi(x) = \frac{1}{\pi} \int\limits_{-\infty}^{+\infty} \frac{e^{i\omega x}}{(1+i\omega)(1-i\omega-\lambda)}\,d\omega \tag{2.13}$$

This is true if (2.11) takes place, i.e.,

$$1 - \frac{\lambda}{1-i\omega} \neq 0$$

or

$$\lambda \neq 1 - i\omega$$

for any $\omega\epsilon(-\infty; +\infty)$. To calculate the integral on the right-hand side of equality (2.13), one can use the theory of residues. Consider the analytic continuation of the integrand from the real axis ω to the complex plane z:

$$\psi(z) = \frac{e^{ixz}}{(1+iz)(1-\lambda-iz)}$$

The singular points of the function $\psi(z)$ are $z_1 = i$, $z_2 = i(\lambda-1)$. These are simple poles.

Consider only the case Re $\lambda < 1$. Then, the singular point z_1 lies in the upper half-plane, and the singular point z_2 lies in the lower half-plane. Consider the following cases.

Let $x > 0$.

Then, the integrand has the form $\psi(z) = \Psi(z)e^{imz}$, where $\Psi(z) \to 0$, as $z \to \infty$. Using Jordan's lemma, the integral can be calculated by the formula

$$\int\limits_{-\infty}^{+\infty} \psi(\omega)\,d\omega = 2\pi i \sum_{k=1}^{n} \underset{z=z_k}{Res}\,\psi(z),$$

where z_k is the singular points of the function $\psi(z)$ in the upper half-plane. Then for $x > 0$, we get

$$\varphi(x) = \frac{1}{\pi} \int\limits_{-\infty}^{+\infty} \psi(\omega)d\omega = 2i \operatorname*{Res}_{z=z_1} \psi(z) = \frac{2e^{-x}}{2-\lambda}.$$

Let $x = 0$.

Then $\psi(z) = O\left(1/z^2\right)$, so the rule for calculating the integral using residues used in the case $x > 0$ is still valid. Therefore, for $x = 0$, the resulting solution is the same.

Let $x < 0$.

Then, to use the Jordan's lemma, one must consider semicircles in the lower half-plane, which gives the formula for calculating the integral with the opposite sign

$$\int\limits_{-\infty}^{+\infty} \psi(\omega)d\omega = -2\pi i \sum_{k=1}^{n} \operatorname*{Res}_{z=z_k} \psi(z),$$

where z_K is the singular points of the function in the lower half-plane. Then for $x < 0$,

$$\varphi(x) = \frac{1}{\pi} \int\limits_{-\infty}^{+\infty} \psi(\omega)d\omega = -2i \operatorname*{Res}_{z=z_2} \psi(z) = \frac{2e^{x(1-\lambda)}}{2-\lambda}.$$

So, the solution of the equation for $\lambda \neq 1 - i\omega$, $\omega \in [-\infty, +\infty]$ is obtained in the form

$$\varphi(x) = \begin{cases} \frac{2}{2-\lambda}e^{-x}, & x \geqslant 0 \\ \frac{2}{2-\lambda}e^{x(1-\lambda)}, & x < 0 \end{cases}$$

We will try now to solve the same example in *Wolfram Mathematica*. Set the initial data.

```
In[1]:= Clear[f,K,φ,F,KK,Φ,λ,x,ω]
        f[x_]=Exp[-Abs[x]];
        K[τ_]=Exp[τ]HeavisideTheta[-τ];
        inteqn=φ[x]==f[x]+λ Integrate[K[x-t]φ[t],{t,-∞,+∞}]

        Integrate::idiv: Integral of E^(-t+x) HeavisideTheta[t-
x]

        φ[t] does not converge on {-∞,∞}.

Out[4]:=φ[x]==e^{-Abs[x]}+λ ∫_{-∞}^{∞} e^{-t+x}HeavisideTheta[t-x]φ[t]dt
```

Defining the kernel, we used the **Heaviside** function, as symbolic calculations for the Fourier transform in *Wolfram Mathematica* work better with it than with **Piecewise**. The Heaviside function is defined by the expression

$$\theta(x) = \begin{cases} 0, \ x < 0 \\ 1, \ x > 0 \end{cases}.$$

For $x = 0$, the function can be redefined by different values, for example, 0, 1 or ½. In *Wolfram Mathematica*, the function is undefined at $x = 0$.

Note also that we received a message that the integral diverges. This means that we will not be able to use the *inteqn* variable, which stores the equation, in other functions such as **DSolve** or **FourierTransform** (the reader can verify this for himself).

We calculate the Fourier transform for $f(x)$ and $K(\tau)$, compose an equation for the image $\Phi(\omega)$ and solve it.

It is worth noting that there is no commonly accepted definition for the Fourier transform: The coefficient in front of the integral and the coefficient in the exponent may differ. *Wolfram Mathematica* provides the ability to customize these coefficients. However, the default values do not match our definition of Fourier transforms, so we use the **FourierParameters** option to set them. For more information on the various representations of the Fourier transform in *Wolfram Mathematica*, see the documentation for the **FourierTransform** function.

```
In [5]:= F[ω_]=FourierTransform[f[x],x,ω,
           FourierParameters->{0, -1}]
         KK[ω_]=FourierTransform[K[τ],τ,ω,
           FourierParameters->{0, -1}]
```

$$Out[5]:=\frac{\sqrt{\frac{2}{\pi}}}{1+\omega^2}$$

$$Out[6]:=-\frac{i}{\sqrt{2\pi}(i+\omega)}$$

```
In [7]:= imageeqn=Φ[ω]==F[ω]+λ Sqrt[2π]KK[ω] Φ[ω]
```

$$Out[7]:=\Phi[\omega]==\frac{\sqrt{\frac{2}{\pi}}}{1+\omega^2}+\frac{i\lambda\Phi[\omega]}{i+\omega}$$

```
In [8]:= imagesol=Solve[imageeqn,Φ[ω]][[1,1]]
```

$$Out[8]:=\Phi[\omega]\rightarrow\frac{\sqrt{\frac{2}{\pi}}}{(-i+\omega)(i-i\lambda+\omega)}$$

Now, we apply the inverse Fourier transform and get the function $\phi(x)$.

```
In [9]:=s=InverseFourierTransform[Φ[ω]/.
         imagesol,ω,x,FourierParameters->{0, -1}]
```

$$\text{Out}[9] := -\frac{1}{-2+\lambda} e^{-x(1+\lambda)}$$

$$\left(2e^{x\lambda}\,\text{HeavisideTheta}[x] + e^{2x}\left(\text{Sign}[x](-1+\text{Sign}[\text{Abs}[1-\text{Re}[\lambda]]])\right)\right.$$

$$\left. + 2\,\text{HeavisideTheta}\!\left[-x\,\text{Sign}[1-\text{Re}[\lambda]]\right]\text{Sign}\,[1-\text{Re}[\lambda]]\right)$$

The expression turned out to be cumbersome and explicitly dependent on x and λ. Let us try to figure out this dependence. First, we split into two subsolutions $\text{Re}\,\lambda < 1$ and $\text{Re}\,\lambda > 1$.

```
In [10]:= s1=FullSimplify[s,Re[λ]<1]
          s2=FullSimplify[s,Re[λ]>1]
```

$$\text{Out}[10] := \frac{2e^{x-x\lambda}\,\text{HeavisideTheta}[-x] + 2e^{-x}\,\text{HeavisideTheta}[x]}{2-\lambda}$$

$$\text{Out}[11] := \frac{2e^{x-x\lambda}\,\text{HeavisideTheta}[-x] + 2e^{-x}\,\text{HeavisideTheta}[x]}{2-\lambda}$$

Further, we turn each of the subsolutions into a Piecewise function, considering different values of the variable x. The Heaviside function for $x = 0$ is extended by the value ½. We finally obtain for $\text{Re}\,\lambda < 1$:

```
In [12]:= sol1=Piecewise[{{Refine[#,x<0],x<0},
          {Refine[#,x>0],x>0},
          {Refine[#,x==0],x==0}}]&@s1/.
          HeavisideTheta[0]->1/2//FullSimplify
```

$$\text{Out}[12] := \begin{cases} -\frac{2e^{-x}}{-2+\lambda} & x \geq 0 \\ -\frac{2e^{x-x\lambda}}{-2+\lambda} & \text{True} \end{cases}$$

For $\text{Re}\,\lambda > 1$:

```
In [13]:= sol2=Piecewise[{{Refine[#,x<0],x<0},
          {Refine[#,x>0],x>0},
          {Refine[#,x==0],x==0}}]&@s2/.
          HeavisideTheta[0]->1/2//FullSimplify
```

$$\text{Out}[13] := \begin{cases} \frac{2e^{-x}-2e^{x-x\lambda}}{2-\lambda} & x > 0 \\ 0 & \text{True} \end{cases}$$

The case $\text{Re}\,\lambda < 1$ coincided with the "manual" solution.

Let us also check the obtained solution by substituting it into the original integral equation. For example, for the case Re λ < 1 for different values of the variable x, we obtain

```
In [14]:= Clear[φ1]
          φ1[x_]:=Evaluate@s1
          FullSimplify[inteqn/.φ->φ1,x<0]
          FullSimplify[inteqn/.φ->φ1,x==0] /.
          HeavisideTheta[0]->1/2
          FullSimplify[inteqn/.φ->φ1,x>0]

Out[16]:=True
Out[17]:=True
Out[18]:=True
```

A similar substitution of the solution in the case of Re λ > 1 also gives the equality.

```
In [19]:= Clear[φ1]
          φ2[x_]:=Evaluate@s1
          FullSimplify[inteqn/.φ->φ2,x<0]
          FullSimplify[inteqn/.φ->φ2,x==0] /.
          HeavisideTheta[0]->1/2
          FullSimplify[inteqn/.φ->φ2,x>0]

Out[21]:=True if Re[λ]> 0
Out[22]:=True if Re[λ]> 0
Out[23]:=True if Re[λ]> 0
```

Answer

$$\varphi(x) = \begin{cases} \frac{2}{2-\lambda}e^{-x}, & x \geqslant 0 \\ \frac{2}{2-\lambda}e^{x(1-\lambda)}, & x < 0 \end{cases}$$

Example 2.13 Solve the equation

$$\int\limits_{-\infty}^{+\infty} \frac{\varphi(t)\mathrm{d}t}{1 + (x - t)^2} = \frac{x}{x^2 + 4}.$$

Solution

This is a Fredholm equation of the first kind with infinite limits of integration. Let us solve it using the Fourier transform. Kernel is

$$K(\tau) = \frac{1}{1 + \tau^2},$$

where $\tau = x - t$. Find its Fourier transform:

$$\tilde{K}(\omega) = \frac{1}{\sqrt{2\pi}} \int\limits_{-\infty}^{+\infty} K(t) e^{-i\omega\tau} dt = \frac{1}{\sqrt{2\pi}} \int\limits_{-\infty}^{+\infty} \frac{e^{-i\omega\tau} d\tau}{1 + \tau^2}.$$

Let us mark

$$I = \int\limits_{-\infty}^{+\infty} \frac{e^{-i\omega\tau} d\tau}{1 + \tau^2}.$$

To calculate the integral I, it is necessary to consider separately the cases $\omega < 0$ and $\omega > 0$.

If $\omega < 0$, then the Jordan's lemma can be used to calculate the integral. Then, $I = 2\pi i \operatorname*{Res}\limits_{z=i} \psi(z)$, where $\psi(z) = \frac{e^{-i\omega z}}{1+z^2}$, and the point $z = i$ is a simple pole of the function $\psi(z)$ lying in the upper half-plane,

$$\operatorname*{Res}_{z=i} \psi(z) = \lim_{z \to i} \frac{e^{-i\omega z}}{1 + z^2}(z - i) = \left. \frac{e^{-i\omega z}}{z + i} \right|_{z=i} = \frac{e^{\omega}}{2i}.$$

Then, if $\omega < 0$

$$I = 2\pi i \frac{e^{\omega}}{2i} = \pi e^{\omega}.$$

If $\omega > 0$, then to calculate the integral, one can use a modification of Jordan's lemma, according to which $I = -2\pi i \operatorname*{Res}\limits_{z=-i} \psi(z)$, where the point $-i$ is a simple pole of $\psi(z)$ in the lower half-plane:

$$\operatorname*{Res}_{z=-i} \psi(z) = \lim_{z \to -i} \frac{e^{-i\omega z}}{1 + z^2}(z + i) = \left. \frac{e^{-i\omega z}}{z - i} \right|_{z=-i} = \frac{e^{-\omega}}{-2i}.$$

Then, if $\omega > 0$,

$$I = -2\pi i \frac{e^{-\omega}}{-2i} = \pi e^{-\omega}.$$

Finally,

$$\tilde{K}(\omega) = \begin{cases} \sqrt{\frac{\pi}{2}} e^{\omega}, & \omega < 0 \\ \sqrt{\frac{\pi}{2}} e^{-\omega}, & \omega > 0 \end{cases}.$$

For the free term $f(x)$, we find the Fourier transform in the same way:

$$F(\omega) = \frac{1}{\sqrt{2\pi}} \int\limits_{-\infty}^{+\infty} f(x)e^{-i\omega x}d\omega = \frac{1}{\sqrt{2\pi}} \int\limits_{-\infty}^{+\infty} \frac{xe^{-i\omega x}}{x^2+4}dx.$$

If $\omega < 0$,

$$F(\omega) = \frac{2\pi i}{\sqrt{2\pi}} \operatorname*{Res}_{z=2i} \frac{ze^{-i\omega z}}{z^2+4} = \sqrt{2\pi}i \lim_{z\to 2i} \frac{ze^{-i\omega z}(z-2i)}{z^2+4} = \frac{\sqrt{2\pi}}{2}ie^{2\omega}$$

If $\omega > 0$,

$$F(\omega) = -\frac{2\pi i}{\sqrt{2\pi}} \operatorname*{Res}_{z=-2i} \frac{ze^{-i\omega z}}{z^2+4} = -\sqrt{2\pi}i \lim_{z\to -2i} \frac{ze^{-i\omega z}(z+2i)}{z^2+4} = -\frac{\sqrt{2\pi}}{2}ie^{-2\omega}.$$

Finally,

$$F(\omega) = \begin{cases} \sqrt{\frac{\pi}{2}}ie^{2\omega}, & \omega < 0 \\ \sqrt{\frac{\pi}{2}}ie^{-2\omega}, & \omega > 0 \end{cases}.$$

Matching the unknown function $\varphi(x)$ with its Fourier transform $\Phi(\omega)$, we obtain the equation in the Fourier images

$$\sqrt{2\pi}\,\widetilde{K}(\omega)\Phi(\omega) = F(\omega),$$

from where

$$\Phi(\omega) = \frac{1}{\sqrt{2\pi}} \frac{F(\omega)}{\widetilde{K}(\omega)} = \begin{cases} \frac{i}{\sqrt{2\pi}}e^{\omega}, & \omega < 0; \\ -\frac{i}{\sqrt{2\pi}}e^{-\omega}, & \omega > 0 \end{cases}.$$

We find the solution of the integral equation using the inverse Fourier transform

$$\varphi(x) = \frac{1}{\sqrt{2\pi}} \int\limits_{-\infty}^{+\infty} \varphi(\omega)e^{\omega x}d\omega = \frac{1}{\sqrt{2\pi}}\left[\int\limits_{-\infty}^{0} \frac{i}{\sqrt{2\pi}}e^{\omega}e^{\omega x}dx - \int\limits_{0}^{+\infty} \frac{i}{\sqrt{2\pi}}e^{-\omega}e^{i\omega x}d\omega\right]$$

$$= \frac{i}{2\pi}\left[\frac{1}{1+ix}e^{\omega(1+ix)}\Big|_{-\infty}^{0} + \frac{1}{4-ix}e^{-\omega(1-ix)}\Big|_{0}^{+\infty}\right] = \frac{x}{\pi(1+x^2)}$$

Here was used that

$$\left|e^{\pm\omega(1\pm ix)}\right| = e^{\pm\omega} \to 0, \quad \omega \to \mp\infty.$$

This example can be solved in *Wolfram Mathematica*. So, we set the initial data and find the solution in the Fourier images in the same way as in the previous example.

```
In[24]:= Clear[f,K,φ,F,KK,Φ,x,ω]
         f[x_]=x/(x^2+4);
         K[τ_]=1/(1+τ^2);
         inteqn=Integrate[K[x-t]φ[t],{t,-∞,+∞}]==f[x]
```

$$\text{Out}[27]:=\int_{-\infty}^{\infty}\frac{\phi[t]}{1+(-t+x)^2}dt==\frac{x}{4+x^2}$$

```
In[28]:= F[ω_]=FourierTransform[f[x],x,ω,
         FourierParameters->{0, -1}]
         KK[ω_]=FourierTransform[K[τ],τ,ω,
         FourierParameters->{0, -1}]
```

$$\text{Out}[28]:=ie^{-2\omega}\sqrt{\frac{\pi}{2}}\left(e^{4\omega}\text{HeavisideTheta}[-\omega]-\text{HeavisideTheta}[\omega]\right)$$

$$\text{Out}[29]:=e^{-\text{Abs}[\omega]}\sqrt{\frac{\pi}{2}}$$

```
In[30]:= imageeqn=λ Sqrt[2π]KK[ω]Φ[ω]==F[ω]
```

$$\text{Out}[30]:=e^{-\text{Abs}[\omega]}\pi\,\Phi[\omega]$$

$$==ie^{-2\omega}\sqrt{\frac{\pi}{2}}\left(e^{4\omega}\text{HeavisideTheta}[-\omega]-\text{HeavisideTheta}[\omega]\right)$$

```
In[31]:= imagesol=Solve[imageeqn,Φ[ω]][[1,1]]
```

$$\text{Out}[31]:=\Phi[\omega]\rightarrow\frac{ie^{-2\omega+\text{Abs}[\omega]}\left(e^{4\omega}\text{HeavisideTheta}[-\omega]-\text{HeavisideTheta}[\omega]\right)}{\sqrt{2\pi}}$$

Find the original by the inverse Fourier transform.

```
In[32]:= sol=InverseFourierTransform[Φ[ω]/.
         imagesol,ω,x,FourierParameters->{0, -1}]
```

$$\text{Out}[32]:=\frac{x}{\pi+\pi x^2}$$

As usual, we will check the solution by substituting it into the original equation.

```
In[33]:= Clear[φ1]
         φ1[x_]:=Evaluate@sol
         inteqn/.φ->φ1
```

$$\text{Out}[35]:=\textbf{True if }-1<\text{Im}[x]<1$$

Since the variable x is assumed to be real, this means true equality.

Answer

$$\varphi(x) = \frac{x}{\pi(1+x^2)}$$

This example shows that using the *Wolfram Mathematica* package can greatly reduce computation compared to manual calculations.

Example 2.14 Solve the equation

$$\int_0^\infty \varphi(t)\cos xt\, dt = f(x),$$

where the free term is defined as

$$f(x) = \begin{cases} \cos x, & 0 \leqslant x \leqslant \pi \\ 0, & x > \pi \end{cases}.$$

Solution
If the left side of the equation is multiplied by $\sqrt{2/\pi}$, then we get the cosine transformation of the function $\varphi(t)$. Thus, the equation has the form

$$\Phi_c(x) = \sqrt{\frac{2}{\pi}} f(x), \tag{2.14}$$

from where

$$\varphi(t) = \sqrt{\frac{2}{\pi}} \int_0^\infty \Phi_c(x)\cos xt\, dx = \frac{2}{\pi} \int_0^\infty f(x)\cos xt\, dx =$$

$$\frac{2}{\pi} \int_0^\pi \cos x \cos xt\, dx = \frac{1}{\pi} \int_0^\pi [\cos x(1+t) + \cos x(1-t)]dx.$$

For further calculation of the integral, two cases must be considered separately:
    If $t \neq 1$,

$$\varphi(t) = \frac{1}{\pi}\left[\frac{1}{1+t}\sin x(1+t) + \frac{1}{1-t}\sin x(1-t) \right]\Big|_0^\pi = \frac{2t\sin\pi t}{\pi(1-t^2)}.$$

If $t = 1$,

$$\varphi(t) = \frac{2}{\pi} \int\limits_0^\pi \cos^2 x \, dx = 1.$$

Finally,

$$\varphi(t) = \begin{cases} \frac{2t \sin \pi t}{\pi(1-t^2)}, & t \neq 1; \\ 1, & t = 1. \end{cases}$$

Note:

$$\lim_{t \to 1} \varphi(t) = \lim_{t \to 1} \frac{2t \sin \pi t}{\pi(1-t^2)} = \lim_{t \to 1} \frac{2t \sin[\pi(t-1) + \pi]}{\pi(1-t)(1+t)}$$

$$= \frac{1}{\pi} \lim_{t \to 1} \frac{-\sin \pi(t-1)}{1-t} = 1,$$

so that the point $t = 1$ is a removable discontinuity point of the solution $\varphi(t)$.

We will solve this example in *Wolfram Mathematica*, namely we will find the inverse Fourier cosine transform for Eq. (2.14). Let us define a function $f(x)$. It is convenient to set it in the form of a Heaviside rectangular pi-function.

```
In [36]:= Clear[f,x,ω]
          f[x_]=Cos[x]HeavisidePi[(x-π/2)/π];
```

Find the inverse cosine transform. There is a built-in function for this.

```
In [38]:= InverseFourierCosTransform[Sqrt[2/π]f[ω],ω,x,
          FourierParameters->{0, -1}]
```

$$\text{Out [38] := } -\frac{2x \sin[\pi x]}{\pi(-1+x^2)}$$

The solution is the same as when calculating the integral manually.

Answer

$$\varphi(t) = \begin{cases} \frac{2t \sin \pi t}{\pi(1-t^2)}, & t \neq 1; \\ 1, & t = 1. \end{cases}$$

References

L.A. Sakhnovich, *Integral Equations with Difference Kernels on Finite Intervals*. (– Basel – Boston: Birkhäuser Verlag, 1996)

V. Volterra, Theory of Functionals and of Integral and Integro-Differential Equations, (Dover Publ., New York, 1959)

G. Gripenberg, S.-O Londen, O. Staffans, *Volterra Integral and Functional Equations*. (– Cambridge – New York: Cambridge Univ. Press, 1990)

I. Sneddon , *The Use of Integral Transforms*, (McGraw-Hill, New York, 1972)

B. Davis, *Integral Transforms and Their Applications*. (- New York: Springer – Verlag, 1978)

J.W. Miles, *Integral Transforms in Applied Mathematics*. (– Cambridge: Cambridge Univ. Press, 1971)

I. Sneddon, *Fourier Transforms*, (Dover Publications, New York, 1995)

E. C Titchmarsh, Theory of Fourier Integrals, (Oxford Univ. Press, Oxford, 1937)

E. C Titchmarsh, *Introduction to the Theory of Fourier Integrals*, 3rd Edition, (Chelsea Publishing, New York, 1986)

M. L. Krasnov, A. I. Kiselev, and G. I Makarenko, *Problems and Exercises in Integral Equations*, (Mir Publ., Moscow, 1971)

Wolfram Mathematica, http://www.wolfram.com/mathematica/

Chapter 3
Fredholm Theory

3.1 Solution of Fredholm Integral Equations by the Resolvent Method: Method of Fredholm Determinants

Solution of the Fredholm equation of the second kind

$$\varphi(x) = f(x) + \lambda \int_a^b K(x,t)\varphi(t)\mathrm{d}t \tag{3.1}$$

can be found as follows. Consider the equation in an operator form

$$\left(\hat{I} - \lambda\hat{K}\right)\varphi = f,$$

where $\hat{K}$ is an integral operator acting in some Hilbert space H and λ is a numerical parameter. Formally, the solution of such an operator equation can be written as

$$\varphi = \left(\hat{I} - \lambda\hat{K}\right)^{-1} f.$$

This formula really determines the solution if $|\lambda| < \hat{K}^{-1}$. Then, the operator $(\hat{I} - \lambda\hat{K})^{-1}$ exists, and is defined and bounded on the whole Hilbert space H.

If $\hat{K}$ is the Hilbert–Schmidt integral operator, that is, the operator determined by a square-integrable kernel $K(x,t)$, then the operator $(\hat{I} - \lambda\hat{K})^{-1}$ for sufficiently small values λ can be written as the sum $\hat{I} + \hat{\Gamma}(\lambda)$ of the unit operator $\hat{I}$ and some integral operator $\hat{\Gamma}(\lambda)$, which is also the Hilbert–Schmidt operator with a square-integrable kernel depending on parameter λ. This statement will be proved in the next section. Then, the solution will take the form

$$\varphi = \left(\hat{I} + \hat{\Gamma}(\lambda) \right) f.$$

The function $\Gamma(x, t, \lambda)$ by which the solution is presented is called the resolvent kernel $K(x, t)$ in the circle $|\lambda| < \hat{K}^{-1}$.

Thus, the solution to the Fredholm integral Eq. (3.1) can be rewritten as

$$\varphi(x) = f(x) + \lambda \int_a^b \Gamma(x, t, \lambda) f(t) \mathrm{d}t. \tag{3.2}$$

Let us generalize the concept of a resolvent kernel. The Fredholm resolvent is a function $\Gamma(x, t, \lambda)$ defined almost everywhere on the Hilbert space H for all regular values of the parameter λ and representing the solution of equation in the form (3.2).

Fredholm showed that for an integral operator $\hat{K}$ defined by a bounded and continuous kernel $K(x, t)$, a solution to Eq. (3.1) can be found in the following way.

Being a meromorphic function, the resolvent can be represented as a quotient of some two analytic functions in terms of λ. In this case, the poles of the resolvent (the characteristic numbers of a kernel) do not depend on x and t; therefore, it is possible to ensure that the denominator of the resolvent depends only on the parameter λ. Thus, the resolvent will take the following form

$$\Gamma(x, t, \lambda) = \frac{D(x, t, \lambda)}{D(\lambda)}, \tag{3.3}$$

if $D(\lambda) \neq 0$, where $D(x, t, \lambda)$ and $D(\lambda)$ are analytic functions of λ. If we succeed in constructing these functions, then the resolvent will become known and the solution of the integral equation can be found by formula (3.2). For functions $D(x, t, \lambda)$ and $D(\lambda)$, Fredholm gave a representation in the form of power series in terms of parameter λ, and we call it as Fredholm series defined below:

$$D(x, t, \lambda) = \sum_{n=0}^{\infty} \frac{(-1)^n}{n!} B_n(x, t) \lambda^n,$$

$$D(\lambda) = \sum_{n=0}^{\infty} \frac{(-1)^n}{n!} C_n \lambda^n. \tag{3.4}$$

The function $D(x, t, \lambda)$ is called the *Fredholm minor*, and the function $D(\lambda)$ is called the *Fredholm determinant*. The coefficients of the power series are determined by the following formula:

$$C_0 = 1; \ B_0(x, t) = K(x, t),$$

$$C_n = \int\limits_a^b B_{n-1}(x, x)\mathrm{d}x,$$ (3.5)

$$B_n(x, t) = \int\limits_a^b \dots \int\limits_a^b \begin{vmatrix} K(x, t) & K(x, t_1) & K(x, t_n) \\ K(t_1, t_1) & K(t_1, t_2) & K(t_1, t_n) \\ K(t_n, t_1) & K(t_n, t_2) & K(t_n, t_n) \end{vmatrix} \mathrm{d}t_1 \dots \mathrm{d}t_n.$$

For coefficients $B_n(x, t)$, as well as for C_n, the following recurrent relations can be used:

$$B_n(x, t) = C_n K(x, t) - n \int\limits_a^b K(x, s) B_{n-1}(s, t)\mathrm{d}s.$$

It can be seen from the recurrence relations (3.5) and formulas (3.4) and (3.3) that the resolvent depends only on the kernel of the equation $K(x, t)$. If $D(\lambda) \neq 0$, then equation has a unique solution expressed by the formula (3.2). In particular, the homogeneous equation for $D(\lambda) \neq 0$ has a unique solution $\varphi(x) \equiv 0$, called a trivial solution. The values of the parameter λ, for which the homogeneous equation has a unique trivial solution, are called *regular*. For regular values $D(\lambda) \neq 0$, there is a unique solution of (3.1) for any free term $f(x)$.

The characteristic values are those values of the parameter λ, for which the homogeneous equation has nontrivial solutions, called *eigenfunctions* of the kernel. For characteristic values $D(\lambda) = 0$, the resolvent does not exist and Eq. (3.1) either has no solution at all or has an infinite set of them depending on the free term.

Thus, the so-called *Fredholm alternative is* valid: either the inhomogeneous Fredholm integral equation of the second kind is solvable for any free term, or the corresponding homogeneous equation has nontrivial solutions.

More details on the relevant theoretical background can be found in [1–7]. Meanwhile, more methods of solving both Fredholm equations and Volterra equations can be found in the handbook [8]. For more examples and exercises, the interested readers can refer to [9]. The following paragraph provides some numerical examples to solve Fredholm integral equations by using Mathematica software.

Example 3.1 Solve the equation using the Fredholm determinants,

$$\varphi(x) = e^{-x} + \lambda \int\limits_0^1 x e^t \varphi(t)\mathrm{d}t.$$

Solution
Find the coefficients of the Fredholm series

$$B_0(x, t) = K(x, t) = xe^t; \ C_0 = 1;$$

$$C_1 = \int_0^1 B_0(t, t)\mathrm{d}t = \int_0^1 te^t\mathrm{d}t = \left(te^t - e^t\right)\Big|_0^1 = 1;$$

$$B_1(x, t) = C_1 K(x, t) - \int_0^1 K(x, z)B_0(z, t)\mathrm{d}z =$$

$$xe^t - \int_0^1 xe^z ze^t\mathrm{d}z = xe^t - xe^t = 0;$$

$$C_2 = \int_0^1 B_1(t, t)\mathrm{d}t = 0;$$

$$B_2(x, t) = 0; \ldots; C_k = 0, k = 2, 3, \ldots;$$

$$B_k(x, t) = 0, k = 1, 2, \ldots$$

Then,

$$D(\lambda) = C_0 - C_1\lambda = 1 - \lambda;$$

$$D(x, t, \lambda) = B_0(x, t) = xe^t;$$

$$R(x, t, \lambda) = \frac{xe^t}{1 - \lambda} \quad \text{if } \lambda \neq 1.$$

If $\lambda \neq 1$, there is a unique solution of the integral equation:

$$\varphi(x) = f(x) + \lambda \int_0^1 R(x, t, \lambda)f(t)\mathrm{d}t,$$

i.e.,

$$\varphi(x) = e^{-x} + \lambda \int_0^1 \frac{xe^t}{1 - \lambda}e^{-t}\mathrm{d}t;$$

$$\varphi(x) = e^{-x} + \frac{\lambda x}{1 - \lambda}.$$

All values $\lambda \neq 1$ are regular. For these values of the parameter λ, the homogeneous equation

$$\varphi(x) = \lambda \int\limits_0^1 xe^t \varphi(t)dt$$

has the unique trivial solution $\varphi(x) = 0$.

For $\lambda = 1$, homogeneous equation has nontrivial solutions. Indeed, let us check that the function $\varphi(x) = x$ is a solution of the homogeneous equation. Substituting $\varphi(x) = x$ into the equation for $\lambda = 1$, we obtain the identity $x \equiv x$. It is obvious that the function $\varphi(x) = Cx$ is a solution of the homogeneous equation for $\lambda = 1$, i.e., $\varphi(x)$ is an eigenfunction of the kernel and C is an arbitrary constant.

This method can be implemented by using the *Wolfram Mathematica* package [10].

We define the functions to implement the recursive calculation of the coefficients $B_n(x, t)$ and C_n.

```
In [1]:=Clear[Bn,Cn]
        Cn[0]=1;
        Cn[n_]:=Integrate[Bn[x,x,n-1],{x,a,b}]
        Bn[x_,t_,0]:=K[x,t]
        Bn[x_,t_,n_]:=Module[{s},
        Cn[n]K[x,t]-n Integrate[K[x,s]Bn[s,t,n-1],{s,a,b}]
        ]
```

And set the initial data.

```
In [6]:=Clear[f,K,λ,a,b]
        f=Function[{x},Exp[-x]];
        K=Function[{x,t},x Exp[t]];
        a=0;
        b=1;
        eqn=ϕ[x]==f[x]+λ Integrate[K[x,t]ϕ[t],{t,a,b}]
```

$$\text{Out[11]} := \phi[x] == e^{-x} + \lambda \int\limits_0^1 e^t x\phi[t]dt$$

Display several coefficients from the Fredholm series together with numbers. Note that starting from $n = 2$, the coefficients turn to be zero.

```
In [12]:=Column@Table[{n,Cn[n],Bn[x,t,n]},{n,0,5}]
```

$$\text{Out}[12] := \left\{0, 1, e^t x\right\}$$
$$\{1, 1, 0\}$$
$$\{2, 0, 0\}$$
$$\{3, 0, 0\}$$
$$\{4, 0, 0\}$$
$$\{5, 0, 0\}$$

Define functions for calculating the Fredholm minor and the Fredholm determinant, as the well as resolvent.

```
In [13]:=Clear[FredholmMinor,FredholmDet,Resolvent]
        FredholmMinor[x_,t_,λ_,n_]:=
        Sum[(-1)^k/k! Bn[x,t,k]λ^k,{k,0,n}]
        FredholmDet[λ_,n_]:=Sum[(-1)^k/k!Cn[k]λ^k,{k,0,n}]
        Resolvent[x_,t_,λ_,n_]:=
        FredholmMinor[x,t,λ,n]/FredholmDet[λ,n]
```

Calculate the resolvent, knowing that the Fredholm coefficients for $n > 1$ are equal to zero, and find a solution for regular values λ.

```
In [17]:= res=Resolvent[x,t,λ,1]
```

$$\text{Out}[17] := \frac{e^t x}{1-\lambda}$$

```
In [18]:=sol=φ->Function[{x},
        f[x]+λ Integrate[res f[t],{t,a,b}]//Evaluate]
```

$$\text{Out}[18] := \phi \rightarrow \textbf{Function}\left[\{x\}, e^{-x} + \frac{x\lambda}{1-\lambda}\right]$$

We can check the solution by substituting it into the original equation.

```
In [19]:= eqn/.sol//FullSimplify
```

$$\text{Out}[19] := \textbf{True}$$

The graphs of the family of solutions for regular values λ, i.e., for $\lambda \neq 1$ are shown in Fig. 3.1.

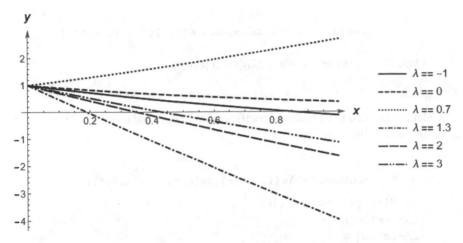

Fig. 3.1 Family of solutions of an integral equation for different values of the parameter λ (Example 3.1)

```
In [20]:=Plot[φ[x]/.sol/.λ->#//Evaluate,{x,a,b},
        PlotLegends->ToString/@Thread
        [λ==#]]&@{-1,0,0.7,1.3,2,3}
```

Answer

$$\varphi(x) = e^{-x} + \frac{\lambda x}{1 - \lambda}, \quad \text{if } \lambda \neq 1.$$

Example 3.2 Solve integral equation

$$\varphi(x) - \lambda \int_0^{2\pi} \operatorname{Sin}(x + t)\varphi(t)\mathrm{d}t = 1.$$

Solution
The solution can be found using the functions defined in the previous example. First, we set the initial data.

```
In [22]:=Clear[f,K,λ,a,b]
        f=Function[{x},1];
        K=Function[{x,t},Sin[x+t]];
        a=0;
        b=2π;
```

```
eqn=φ[x]==f[x]+λ Integrate[K[x,t]φ[t],{t,a,b}]]
```

$$\text{Out[27]} := \phi[x] == 1 + \lambda \int_0^{2\pi} \text{Sin}[t + x]\phi[t]dt$$

Calculate the first few Fredholm coefficients to understand to what order we need to define the Fredholm series.

```
In [28]:=Column@Table[{n,Cn[n],Bn[x,t,n]},{n,0,5}]
```

$\text{Out[28]} := \{0, 1, \text{Sin}[t + x]\}$
$\{1, 0, -\pi \text{Cos}[t - x]\}$
$\{2, -2\pi^2, 0\}$
$\{3, 0, 0\}$
$\{4, 0, 0\}$
$\{5, 0, 0\}$

It can be seen that the coefficients turn to be zero starting from $n = 3$. Now, we calculate the resolvent and find the solution for regular values $\lambda \neq \pm\frac{1}{\pi}$.

```
In [29]:=res=Resolvent[x,t,λ,2]
```

$$\text{Out[29]} := \frac{\pi\lambda\text{Cos}[t-x]+\text{Sin}[t+x]}{1-\pi^2\lambda^2}$$

```
In [30]:=sol=φ->Function[{x},
         f[x]+λ Integrate[resf[t],{t,a,b}]//Evaluate]
```

$\text{Out[30]} := \phi \to \text{Function}[\{x\}, 1]$

Let us check the solution by substitution into the equation.

```
In [31]:= eqn/.sol//FullSimplify
```

$\text{Out[31]} := \textbf{True}$

Answer

$$\varphi(x) = 1, \quad \text{for regular values} \quad \lambda \neq \pm\frac{1}{\pi}.$$

3.2 Iterated Kernels Method

Consider the Fredholm equation of the second kind

$$\varphi(x) = \lambda \int_a^b K(x,t)\varphi(t)\,dt + f(x). \tag{3.6}$$

Equation (3.6) can be rewritten in the operator notation form

$$\left(\hat{I} - \lambda\hat{K}\right)\varphi = f,$$

where $\hat{K}$ is an integral operator acting in some Hilbert space H and λ is a numerical parameter, and we write its formal solution as

$$\varphi = \left(\hat{I} - \lambda\hat{K}\right)^{-1} f. \tag{3.7}$$

If $|\lambda| < \hat{K}^{-1}$, then the operator $\left(\hat{I} - \lambda\hat{K}\right)^{-1}$ exists, is defined on the entire Hilbert space H and is bounded. It can be represented as the sum of the power series:

$$\left(\hat{I} - \lambda\hat{K}\right)^{-1} = \hat{I} + \lambda\hat{K} + \lambda^2\hat{K}^2 + \cdots + \lambda^n\hat{K}^n + \cdots .$$

Therefore, solution (3.7) of Eq. (3.6) can be written as [4, 5]

$$\varphi = f + \lambda\hat{K}f + \lambda^2\hat{K}^2 f + \cdots + \lambda^n\hat{K}^n f + \cdots . \tag{3.8}$$

The same result will be obtained if we will find a solution of Eq. (3.6) in the form of a power series:

$$\varphi_\lambda = \varphi_0 + \lambda\varphi_1 + \cdots + \lambda^n\varphi_n + \cdots$$

(where φ_i does not depend on λ). Substituting this series into Eq. (3.6) and then equating the coefficients of the same degrees of the parameter λ in both sides of the equality, we get:

$$\varphi_0 + \lambda\varphi_1 + \cdots + \lambda^n\varphi_n + \cdots = f + \lambda\hat{K}\left(\varphi_0 + \lambda\varphi_1 + \cdots + \lambda^n\varphi_n + \cdots\right)$$

and

$$\varphi_0 = f, \varphi_1 = \hat{K}\varphi_0 = \hat{K}f, \ldots, \varphi_n = \hat{K}\varphi_{n-1} = \hat{K}^n f, \ldots,$$

i.e., series (3.8).

Let us show that if $\hat{K}$ is a Hilbert–Schmidt integral operator, that is, an operator determined by a square-integrable kernel $K(x, t)$, then the operator $\left(\hat{I} - \lambda\hat{K}\right)^{-1}$ for sufficiently small values of the parameter λ can be written as the sum $\hat{I} + \hat{\Gamma}()$ of the unit operator $\hat{I}$ and some Hilbert–Schmidt integral operator $\lambda\hat{\Gamma}(\lambda)$ with a square-integrable kernel depending on the parameter λ. Let us firstly clarify how the kernels of operators $\hat{K}^2$, $\hat{K}^3$, etc. are written. Consider a more general question.

Two integral operators are given as $\hat{R}_1$ and $\hat{R}_2$:

$$\left(\hat{R}_i\varphi\right)(x) = \int_a^b R_i(x, t)\varphi(t)\mathrm{d}t, \quad i = 1, 2$$

where

$$\int_a^b \int_a^b |R_i(x, t)|^2 \mathrm{d}x\mathrm{d}t = r_i^2 < \infty.$$

To find the kernel of the operator $\hat{R}_1\hat{R}_2$, we write:

$$\left(\hat{R}_1\hat{R}_2\varphi\right)(x) = \int_a^b R_1(x, s)\left[\int_a^b R_2(s, t)\varphi(t)\mathrm{d}t\right]\mathrm{d}s$$

$$= \int_a^b \left[\int_a^b R_1(x, s)R_2(s, t)\mathrm{d}s\right]\varphi(t)\mathrm{d}t.$$

The possibility of changing the order of integration here follows from the Fubini's theorem, since the integrand $R_1(x, s)R_2(s, t)\varphi(t)$ is summable over a set of variables s and t as a product of two functions $R_1(x, s)\varphi(t)$ and $R_2(s, t)$, summable with square.

Consider

$$R(x, t) = \int_a^b R_1(x, s)R_2(s, t)\mathrm{d}s; \tag{3.9}$$

by the Cauchy–Bunyakovsky inequality, we have

$$|R(x, t)|^2 \leq \int_a^b |R_1(x, s)|^2\mathrm{d}s \int_a^b |R_2(x, s)|^2\mathrm{d}s,$$

hence,

$$\int\limits_a^b \int\limits_a^b |R(x,t)|^2 \mathrm{d}x\mathrm{d}t = r_1^2 r_2^2 < \infty.$$

Thus, the product of two integral operators of the Hilbert–Schmidt type is an operator of the same type, with the kernel defined by formula (3.9). In particular, setting $R_1 = R_2 = K$, we obtain that $\hat{K}^2$ is an integral operator with kernel

$$K_2(x,t) = \int\limits_a^b K(x,s)K(s,t)\mathrm{d}s,$$

which satisfies the following condition

$$\int\limits_a^b \int\limits_a^b |K_2(x,t)|^2 \mathrm{d}x\mathrm{d}t \le \left[\int\limits_a^b \int\limits_a^b |K(x,t)|^2 \mathrm{d}x\mathrm{d}t \right]^{-2} = k^4 < \infty,$$

from where $\hat{K}^2 \le k^2$, with

$$k^2 = \int\limits_a^b \int\limits_a^b |K(x,t)|^2 \mathrm{d}x\mathrm{d}t.$$

Similarly, we find that each of the operators $\hat{K}_n$ is determined by the kernel, which can be found using the recurrence relation

$$K_n(x,t) = \int\limits_a^b K_{n-1}(x,s)K(s,t)\mathrm{d}s, \quad n = 2,3,\ldots,$$

$$K_1(x,t) = K(x,t),$$

$$\tag{3.10}$$

satisfying the condition

$$\int\limits_a^b \int\limits_a^b |K_n(x,t)|^2 \mathrm{d}x\mathrm{d}t \le k^{2n}. \tag{3.11}$$

Kernels $K_n(x,t)$ are called *iterated kernels*. If $|\lambda| < k^{-1}$, then the series

$$K(x,t) + \lambda K_2(x,t) + \cdots + \lambda^{n-1} K_n(x,t) + \cdots$$

converge by estimating (3.11) in space $L_2([a, b][a, b])$ to some function $\Gamma(x, t, \lambda)$, summable with square over x and t for each $|\lambda| < k^{-1}$. The integral operator $\hat{\Gamma}(\ )$ with kernel $\Gamma(x, t, \lambda)$ is the sum of the converging series

$$\hat{K} + \lambda \hat{K}^2 + \cdots + \lambda^{n-1} \hat{K}^n + \cdots . \tag{3.12}$$

Multiplying this sum by λ and adding the unit operator $\hat{I}$ to it, we get the operator $\left(\hat{I} - \lambda \hat{K}\right)^{-1}$. So, indeed, for $|\lambda| < k^{-1}$, operator $\left(\hat{I} - \lambda \hat{K}\right)^{-1}$ is the sum of the unit operator $\hat{I}$ and the integral operator $\hat{\Gamma}(\lambda)$ multiplied by λ with the kernel

$$\Gamma(x, t, \lambda) = \sum_{n=1}^{\infty} \lambda^{n-1} K_n(x, t). \tag{3.13}$$

The function $\Gamma(x, t, \lambda)$ which represents the solution of Eq. (3.6) is the kernel resolvent of this equation in a circle $|\lambda| < k^{-1}$.

Hence, the solution to the integral Eq. (3.6), written in the form of operator series (3.8), is as follows:

$$\varphi(x) = f(x) + \lambda \int_a^b \sum_{n=1}^{\infty} \lambda^{n-1} K_n(x, t) f(t) dt$$

or

$$\varphi(x) = f(x) + \lambda \int_a^b \Gamma(x, t, \lambda) f(t) dt. \tag{3.14}$$

The series on the right of (3.13) is called the *Neumann series* of the kernel $K(x, t)$.

The condition $|\lambda| < k^{-1}$ is sufficient for the convergence of the series (3.12), but it is not at all necessary. In some cases, this series may be convergent even for all values of the parameter λ. For example, if $\hat{K}$ is a Volterra-type operator with a kernel satisfying the condition $|K(x, t)|\lambda \leq M$, then, as direct calculation shows, for iterated kernels $K_n(x, t)$, the following estimate is valid:

$$|K_n(x, t)| \leq \frac{M^n (b - a)^{n-1}}{(n - 1)!},$$

whence the convergence of series (3.12) and (3.13) follows for any value of the parameter λ.

Consider the Volterra integral equation of the second kind

$$\varphi(x) = \lambda \int\limits_{\alpha}^{x} K(x, t)\varphi(t)\mathrm{d}t + f(x).$$

Its solution can also be expressed by the resolvent

$$\varphi(x) = \int\limits_{a}^{x} \Gamma(x, t, \lambda) f(t)\mathrm{d}t + f(x), \tag{3.15}$$

where the resolvent is the sum of the Neumann series (3.13).

In contrast to the Fredholm equation, the Neumann series for the Volterra equation converges for all values of the parameter λ, and therefore, for all λ, there is a resolvent and a unique solution to the equation in the form (3.15). Thus, for the Volterra equation, all values of the parameter λ are regular.

Iterated kernels $K_n(x, t)$ can be found by the recurrence relation

$$K_n(x, t) = \int\limits_{t}^{x} K_{n-1}(x, s)K(s, t)\mathrm{d}s, n = 2, 3, \ldots,$$

$$K_1(x, t) = K(x, t) \tag{3.16}$$

It is important to note that in contrast to the recurrence relation for the Fredholm Eq. (3.10) in formula (3.16) for the Volterra equation, the lower limit of integration is equal to t, not a.

Example 3.3 Using the iterated kernels, solve the Fredholm equation

$$\varphi(x) = \lambda \int\limits_{0}^{1} xt\varphi(t)\mathrm{d}t + x.$$

Solution
The iterated kernels can be found by the formulas (3.10):

$$K_1(x, t) = K(x, t) = xt,$$

$$K_2(x, t) = \int\limits_{0}^{1} xzzt\mathrm{d}z = \frac{xt}{3},$$

$$K_3(x, t) = \int\limits_{0}^{1} xz \,\frac{zt}{3}\mathrm{d}z = \frac{xt}{3^2}.$$

Obviously, that

$$K_4(x, t) = \frac{xt}{3^3}.$$

Generally,

$$K_m(x, t) = \frac{xt}{3^{m-1}}, \, m = 1, 2, \ldots$$

Then,

$$\Gamma(x, t, \lambda) = \sum_{m=1}^{\infty} \frac{xt}{3^{m-1}} \lambda^{m-1}$$

$$= xt \sum_{m=1}^{\infty} \frac{xt}{3^{m-1}} \left(\frac{\lambda}{3}\right)^{m-1} = xt \frac{1}{1 - \frac{\lambda}{3}}, \quad \text{if} \quad \left|\frac{\lambda}{3}\right| < 1.$$

Here, the formula for the sum of the terms of a geometric progression with the common ratio $\frac{\lambda}{3}$ is used. The solution looks like

$$\varphi(x) = \lambda \int_0^1 xt \frac{1}{1 - \frac{\lambda}{3}} t \mathrm{d}t + x = \frac{\lambda x}{3 - \lambda} + x = \frac{3x}{3 - \lambda}.$$

We get a solution in the form

$$\varphi(x) = \frac{3x}{3 - \lambda}$$

if $|\lambda| < 3$. In fact, the resolvent, and hence the unique solution of the equation, exists for all regular values of the parameter $\lambda \neq 3$.

This example can be solved in *Wolfram Mathematica*. First, we set the initial data.

```
In [1]:=Clear[f,Kernel,λ,a,b]
        f=Function[{x},x];
        Kernel=Function[{x,t},x t];
        a=0;
        b=1;
        eqn=ϕ[x]==f[x]+λ Integrate[Kernel
        [x,t]ϕ[t],{t,a,b}]
```

$$\text{Out[6]} := \phi[x] == x + \lambda \int_0^1 tx\phi[t]\mathrm{d}t$$

Then, we define a function for calculating iterated kernels and display the first few kernels.

```
In [7]:=Clear[Kn]
        KnFred[x_,t_,1]:=Kernel[x,t]
        KnFred[x_,t_,n_/;n>1]:=Module[{s},
        Integrate[KnFred[x,s,n-1]Kernel[s,t],{s,a,b}]
        ]
In[10]:=Column@Table[{n,KnFred[x,t,n]},{n,1,5}]
```

$\text{Out[10]} := \{1, tx\}$

$\left\{2, \frac{tx}{3}\right\}$

$\left\{3, \frac{tx}{9}\right\}$

$\left\{4, \frac{tx}{27}\right\}$

$\left\{5, \frac{tx}{81}\right\}$

Unfortunately, it is not possible to find the sum of a series from the recursive function automatically as shown below:

```
In [11]:=Sum[KnFred[x,t,n] λ^(n-1),{n,1,∞}]
```

$$\text{Out[11]} := \sum_{n=1}^{\infty} \lambda^{-1+n} \text{Kn}[x, t, n]$$

However, if we can get an explicit form for a sequence of iterated kernels, then the sum of the series can be calculated. To do this, we can use the **FindSequenceFunction** by passing the first few members of the sequence to it.

```
In [12]:=seq=FindSequenceFunction
        [Table[KnFred[x,t,k],{k,1,5}],n]
        res=Sum[seq λ^(n-1),{n,1,∞}]
```

$\text{Out[12]} := 3^{1-n} tx$

$\text{Out[13]} := -\frac{3tx}{-3+\lambda}$

Thus, we manage to find the resolvent kernel. We obtain a solution for regular values $\lambda \neq 3$ and check it by substituting it into the original equation.

```
In [14]:=sol=φ->Function[{x},
        f[x]+λ Integrate[res f[t],{t,a,b}]//Evaluate]
```

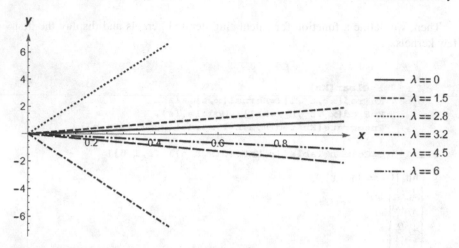

Fig. 3.2 Family of solutions of the integral equation for different values of the parameter λ (Example 3.3)

Out[14] := $\phi \rightarrow$ **Function**$\left[\{x\}, x - \frac{x\lambda}{-3+\lambda}\right]$

In [15]:=**eqn/.sol//FullSimplify**

Out[15] := **True**

The following codes can be used to plot the graphs of solutions for regular values of the parameter λ (Fig. 3.2).

```
In [16]:=optplot={AxesStyle->Arrowheads[{0.0,0.025}],
    AxesLabel->{Style[x,Bold,Medium],Style
    [y,Bold,Medium]}};
    Plot[ϕ[x]/.sol/.λ->#//Evaluate,{x,a,b},
    PlotLegends->ToString/@Thread[λ==#],
    Evaluate@optplot]&@
    {0,1.5,2.8,3.2,4.5,6}
```

Answer

$$\varphi(x) = \frac{3x}{3 - \lambda}, \quad \text{if} \ \ \lambda \neq 3.$$

Example 3.4 Solve the Fredholm equation using the method of iterated kernels

$$\varphi(x) - \lambda \int\limits_{0}^{2\pi} \mathrm{Sin}(x + t)\varphi(t)\mathrm{d}t = 1.$$

Solution

Let us compare the method of Fredholm determinants and the method of iterated kernels. For this, we solve the Fredholm equation from Example 3.2 by the method of iterated kernels.

We use the algorithm given in the previous example. So, we set the initial data and display the first few iterated kernels.

```
In [18]:=Clear[f,Kernel,λ,a,b]
        f=Function[{x},1];
        Kernel=Function[{x,t},Sin[x+t]];
        a=0;
        b=2π;
        eqn=ϕ[x]==f[x]+λ Integrate
        [Kernel[x,t]ϕ[t],{t,a,b}]]
```

Out[23] := $\phi[x] == 1 + \lambda \int\limits_{0}^{2\pi} \mathrm{Sin}[t + x]\phi[t]\mathrm{d}t$

```
In [24]:=Column@Table[{n,KnFred[x,t,n]},{n,1,5}]
```

Out[24] := $\{1, \mathrm{Sin}[t + x]\}$
$\{2, \pi\,\mathrm{Cos}[t - x]\}$
$\{3, \pi^2\mathrm{Sin}[t + x]\}$
$\{4, \pi^3\mathrm{Cos}[t - x]\}$
$\{5, \pi^4\mathrm{Sin}[t + x]\}$

Try to get the law for the sequence of kernels.

```
In [25]:=seq=FindSequenceFunction
        [Table[KnFred[x,t,k],{k,1,5}],n]
```

Out[25] := FindSequenceFunction

$\left[\{\mathrm{Sin}[t+x], \pi\,\mathrm{Cos}[t-x], \pi^2\mathrm{Sin}[t+x], \pi^3\mathrm{Cos}[t-x], \pi^4\mathrm{Sin}[t+x]\}, n\right]$

Sometimes, this cannot be done immediately. So, we can pass more sequence members to the **FindSequenceFunction**.

```
In [26]:=seq=FindSequenceFunction
         [Table[KnFred[x,t,k],{k,1,10}],n]
```

$$\text{Out}[26] := \frac{1}{2}\pi^{-1+n}$$

$$\left(\text{Cos}[t-x]+(-1)^n\text{Cos}[t-x]+\text{Sin}[t+x]+(-1)^{1+n}\text{Sin}[t+x]\right)$$

Now, the rule has been found. This means that the resolvent can be calculated.

```
In [27]:=res=Sum[seq λ^(n-1),{n,1,∞}]
```

$$\text{Out}[27] := -\frac{\pi^2\lambda^2\text{Cos}[t-x]+\pi\lambda\text{Sin}[t+x]}{\pi\lambda(-1+\pi\lambda)(1+\pi\lambda)}$$

The form of the resolvent is somewhat different than the one obtained in Example 3.2; however, using algebraic transformations, it is easy to verify that this is the same function, which exists if $\lambda \neq \pm\frac{1}{\pi}$. We will check this fact by finding a solution of the equation.

```
In[28]:=sol=ϕ->Function[{x},
        f[x]+λ Integrate[res f[t],{t,a,b}]//Evaluate]
```

$$\text{Out}[28] := \phi \rightarrow \text{Function}[\{x\}, 1]$$

The solution turned out to be exactly the same as in Example 3.2.

Answer

$$\varphi(x) = 1, \quad \text{if} \quad \lambda \neq \pm\frac{1}{\pi}.$$

Example 3.5 Solve the Volterra equation using the method of iterated kernels

$$\varphi(x) = \lambda \int\limits_0^x e^{x^2-t^2}\varphi(t)\mathrm{d}t + e^{x^2}$$

Solution
The iterated kernels can be found by formulas (3.16).

$$K_1(x,t) = K(x,t) = e^{x^2 - t^2},$$

$$K_2(x,t) = \int_t^x e^{x^2 - z^2} e^{z^2 - t^2} \, dz = e^{x^2 - t^2}(x - t),$$

$$K_3(x,t) = \int_t^x e^{x^2 - z^2} e^{z^2 - t^2}(z - t) \, dz = e^{x^2 - t^2} \frac{(x - t)^2}{2},$$

$$K_4(x,t) = e^{x^2 - t^2} \frac{(x - t)^3}{3!}.$$

It is clear that

$$K_m(x,t) = e^{x^2 - t^2} \frac{(x - t)^{m-1}}{(m - 1)!}, \, m = 1, 2, 3, \ldots$$

Then,

$$\Gamma(x,t,\lambda) = \sum_{m=1}^{\infty} K_m(x,t) \lambda^{m-1} = \sum_{m=1}^{\infty} e^{x^2 - t^2} \frac{(x - t)^{m-1}}{(m - 1)!} \lambda^{m-1}$$

$$= e^{x^2 - t^2} \sum_{m=1}^{\infty} \frac{[\lambda(x - t)]^{m-1}}{(m - 1)!} = e^{x^2 - t^2} e^{\lambda(x - t)}.$$

Here, we used the well-known expansion of the exponent $e^{\lambda(x-t)}$, which is valid for any values of parameter λ.

The unique solution of the equation is

$$\varphi(x) = \lambda \int_0^x e^{x^2 - t^2} e^{\lambda(x-t)} e^{t^2} \, dt + e^{x^2} = e^{x^2} + \lambda e^{x^2} e^{\lambda x} \int_0^x e^{-\lambda t} \, dt = e^{x^2 + \lambda x}.$$

So,

$$\varphi(x) = e^{x^2 + \lambda x}.$$

This example can be solved in *Wolfram Mathematica*. We set the initial data.

```
In [29]:=Clear[f,Kernel,λ,a,b]
        f=Function[{x},Exp[x^2]];
        Kernel=Function[{x,t},Exp[x^2-t^2]];
        a=0;
```

```
        b=x;
        eqn=φ[x]==f[x]+λ Integrate
        [Kernel[x,t]φ[t],{t,a,b}]
```

$$\text{Out[34]} := \phi[x] == e^{x^2} + \lambda \int_0^x e^{-t^2+x^2}\phi[t]\,dt$$

Define the function for calculating the iterated kernels in the case of the Volterra equation and derive the first few kernels.

```
    In [35]:=Clear[Kn]
            KnVolt[x_,t_,1]:=Kernel[x,t]
            KnVolt[x_,t_,n_/;n>1]:=Module[{s},
            Integrate[KnVolt[x,s,n-1]Kernel[s,t],{s,a,b}]
            ]
    In[38]:=Column@Table[{n,KnVolt[x,t,n]},{n,1,5}]
```

$$\text{Out[38]} := \{1, tx\}$$
$$\left\{2, \frac{tx}{3}\right\}$$
$$\left\{3, \frac{tx}{9}\right\}$$
$$\left\{4, \frac{tx}{27}\right\}$$
$$\left\{5, \frac{tx}{81}\right\}$$

Next, we obtain the formula for the sequence of iterated kernels and find the resolvent. By the form of the resolvent, we can conclude that all values of the parameter λ are regular.

```
    In [39]:=seq=FindSequenceFunction[Table[KnVolt[x,t,k],
            {k,1,5}],n]//FullSimplify
            res=Sum[seq λ^(n-1),{n,1,∞}]
```

$$\text{Out[39]} := \frac{e^{-t^2+x^2}(-t+x)^{-1+n}}{\text{Gamma}[n]}$$
$$\text{Out[40]} := e^{-t^2+x^2-t\lambda+x\lambda}$$

After that, what remains to do is to find the solution and check it by substitution.

```
    In [41]:=sol=φ->Function[{x},
            f[x]+λ Integrate[res f[t],{t,a,b}]//Evaluate]
```

$$\text{Out[41]} := \phi \to \textbf{Function}\left[\{x\}, e^{x^2} + e^{x^2}\left(-1 + e^{x\lambda}\right)\right]$$

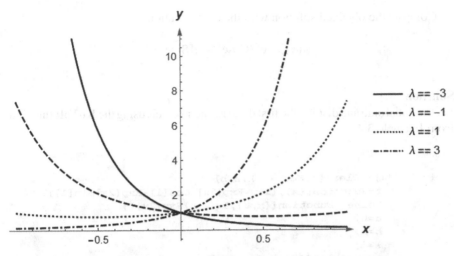

Fig. 3.3 Family of solutions of the integral equation for different values of the parameter λ (Example 3.5)

```
In [42]:=eqn/.sol//FullSimplify
Out[42]:=True
```

Let us plot the graphs of solutions for regular values of the parameter λ (Fig. 3.3).

```
In [43]:=Plot[φ[x]/.sol/.λ->#//Evaluate,{x,a-1,a+1},
        PlotLegends->ToString/@Thread
        [λ==#],Evaluate@optplot]&@
        Range[-3,3]
```

Answer

$$\varphi(x) = e^{x^2 + \lambda x}.$$

Example 3.6 Find an approximate solution of the Volterra integral equation using iterated kernels.

$$\varphi(x) = \left(1 - xe^{2x}\right)\text{Cos}1 - e^{2x}\text{Sin}1 + \int_0^x \left(1 - (x - t)e^{2x}\right)\varphi(t)dt.$$

Compare the obtained solution with the exact solution

$$\varphi(x) = e^x \left(\text{Cose}^x - e^x \text{Sine}^x \right).$$

Solution

Set the initial data and display the first three iterated kernels using the **KnVolt** function from Example 3.4.

```
In [44]:=Clear[f,Kernel,λ,a,b]
        f=Function[{x},(1-x Exp[2x])Cos[1]-Exp[2x]Sin[1]];
        Kernel=Function[{x,t},1-(x-t)Exp[2x]];
        a=0;
        b=x;
        λ=1;
        eqn=φ[x]==f[x]+λ Integrate
        [Kernel[x,t]φ[t],{t,a,b}]
```

$$\text{Out[50]} := \phi[x] == 1 + \lambda \int_0^{2\pi} \text{Sin}[t + x]\phi[t]dt$$

```
In [51]:= Column@Table[{n,KnVolt[x,t,n]},{n,1,5}]
```

$$\text{Out[51]} := \left\{ 1, 1 + e^{2x}(t - x) \right\}$$

$$\left\{ 2, \frac{1}{4}\left(-e^{2t} - 4t + 4x + e^{4x}(-1 - t + x) \right) \right.$$

$$\left. +e^{2x}\left(1 + 2t - 2(t - x)^2 - 2x + e^{2t}(1 - t + x) \right) \right\}$$

$$\left\{ 3, \frac{1}{384}\left(6\left(e^{4t} + 8e^{2t}(-1 + 2t - 2x) + 32(t - x)^2 \right) \right. \right.$$

$$+ 3e^{6x}(3 + 2t - 2x) + 6e^{4x}$$

$$\left(-5 + 8t^2 + 4t\left(1 + e^{2t} - 4x \right) - 4\left(1 + e^{2t} \right)x + 8x^2 \right)$$

$$+ e^{2x}48 + e^{4t}(-9 + 6t - 6x) + 32(-3 + 2t - 2x)(t - x)^2$$

$$\left. \left. +24e^{2t}\left(1 + 2t^2 - 4t(1 + x) + 2x(2 + x) \right) \right) \right) \right\}$$

With more iterations, the kernels found to be in more complex form. The **Find-SequenceFunction** in this case will not help to find the law for the sequence of kernels, as the readers can see for themselves. Therefore, we will find an approximate expression for the resolvent using only the first three iterated kernels.

```
In [52]:=res=Sum[KnVolt[x,t,n] λ^(n-1),{n,1,3}]//FullSimplify
```

$$\textbf{Out}[52] := \frac{1}{384} 3e^{6x}(3 + 2t - 2x)$$

$$\left(+6e^{4x}\left(-21 + 8t^2 + 4t(-3 + e^{2t} - 4x) + 4x(3 - e^{2t} + 2x)\right)\right.$$

$$+6(e^{4t} + 8e^{2t}(-3 + 2t - 2x) + 32(2 + t^2 - 2t(1 + x) + x(2 + x)))$$

$$+ e^{2x}\left(e^{4t}(-9 + 6t - 6x) + 144(1 + 4t - 4x) + 32(-9 + 2t - 2x)\right.$$

$$\left.\left.(t - x)^2 + 24e^{2t}(5 + 2t^2 - 4t(2 + x) + 2x(4 + x))\right)\right)$$

Now, we construct solutions of the equation, which also is an approximate solution. Its form is also quite cumbersome.

```
In [53]:=sol=φ->Function[{x},
         f[x]+λ Integrate[res f[t],{t,a,b}]//Evaluate]
```

$$\textbf{Out}[53]: = \phi \rightarrow \textbf{Function}[\{x\}, (1 - e^{2x}x)\text{Cos}[1] - e^{2x}\text{Sin}[1]$$

$$+\frac{1}{13824}(12(-473 + 6x(123 + 8x(9 + 4x)))\text{Cos}[1]$$

$$+ 36(295 + 24x(11 + 4x))\text{Sin}[1]$$

$$+ e^{8x}((-11 + 6x)\text{Cos}[1] + 6\text{Sin}[1])$$

$$- 3e^{6x}((-119 + 12x(1 + 3x))\text{Cos}[1] + 6(17 + 6x)\text{Sin}[1])$$

$$+ 9e^{4x}((-609 + 2x(111 + 16x(3 + 2x)))$$

$$\text{Cos}[1] + 6(117 + 4x(9 + 4x))\text{Sin}[1])$$

$$+ e^{2x}((10811 - 12x(-104 + 3x(195 + 16x(5 + x))))$$

$$\text{Cos}[1] - 6(2773 + 6x(427 + 4x(45 + 8x)))\text{Sin}[1]))]$$

Also, set the exact solution of the integral equation and check it by substituting into the equation.

```
In [54]:=exactsol=φ->Function[{x},
       Exp[x](Cos[Exp[x]]-Exp[x]Sin[Exp[x]])];
       eqn/.exactsol//FullSimplify
Out[55]:=True
```

To compare the approximate solution with the exact one, let us plot their graphs (Fig. 3.4).

```
In [56]:=Plot[{φ[x]/.sol,φ[x]/.exactsol},{x,a-1.9,a+1.9},
```

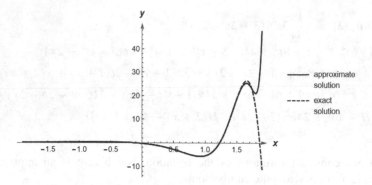

Fig. 3.4 Comparison of the solution to the Volterra equation obtained using iterated kernels (three terms in the expansion for the resolvent) with the exact solution (Example 3.6)

```
PlotLegends->{"approximatesolurion",
"exact solution"},
PlotRange->All,Evaluate@optplot]
```

It can be seen that the "iterated kernels" solution well approximates the exact one in the vicinity of the point $x = 0$ and begins to move away from the exact one with the distance from this point. If more accuracy is required, then more terms in the expansion for the resolvent can be taken. For example, in Fig. 3.5, a solution is constructed when four terms are taken in the expansion for the resolvent (the graph is plotted in the same range along the abscissa).

Answer
The accuracy of the obtained solution strongly depends on the number of terms in the expansion of the series.

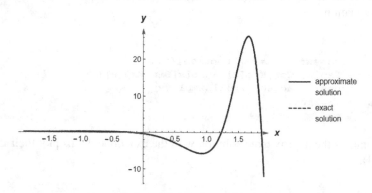

Fig. 3.5 Comparison of the solution to the Volterra equation obtained using iterated kernels (four terms in the expansion for the resolvent) with the exact solution (Example 3.6)

3.3 Characteristic Numbers and Eigenfunctions. Solution of Homogeneous Fredholm Integral Equations with Degenerate Kernel

Homogeneous Fredholm integral equation of the second kind

$$\varphi(x) - \lambda \int_a^b K(x, t)\varphi(t)\mathrm{d}t = 0 \tag{3.17}$$

always has an obvious solution $\varphi(x) \equiv 0$ called a trivial solution. The values of the parameter λ, for which this equation has nonzero solutions $\varphi(x) \neq 0$, are called the characteristic numbers of the Fredholm equation of the second kind or *characteristic numbers* of the kernel $K(x, t)$. Each nonzero solution of this equation is called the *eigenfunction* corresponding to the characteristic number λ. Equation (3.17) has either a finite or a countable set of characteristic numbers; if this set is countable, then it has a unique limit point at infinity.

The number $\lambda = 0$ is not a characteristic number, because from the equation follows that then $\varphi(x) \equiv 0$. Noncharacteristic numbers are called *regular*.

If the kernel $K(x, t)$ is continuous or square summable in the square $[a, b] \times [a, b]$, then each characteristic number λ corresponds to a finite number of linearly independent eigenfunctions; the number p of such functions is called the *rank* of the characteristic number. Each characteristic number of Eq. (3.17) has a finite rank. Different characteristic numbers can have different ranks.

The eigenfunctions are determined up to a constant factor, i.e., if $\varphi(x)$ is an eigenfunction corresponding to some characteristic number λ, then $C\varphi(x)$, where C is an arbitrary constant, is also an eigenfunction corresponding to the same characteristic number λ.

The characteristic numbers are zeros of the Fredholm determinant $D(\lambda)$, i.e., poles of the resolvent $\Gamma(x, t, \lambda)$. However, for some classes of the kernels, it is possible not to construct a function $D(\lambda)$, but to search for characteristic numbers in simpler ways. Degenerate kernels are one of these classes. More about characteristic numbers can be found in [5].

The kernel $K(x, t)$ is called *degenerate*, if it can be represented as a finite sum of products of two factors, one of which depends only on x and the other only on t

$$K(x, t) = \sum_{k=1}^{n} a_k(x)b_k(t). \tag{3.18}$$

In this case, the functions $a_k(x)$ and $b_k(t)$ will be assumed to be continuous in the square $[a, b] \times [a, b]$ and linearly independent from each other.

The following functions are examples of the degenerate kernels:

$$K(x, t) = e^{x-t} = e^x e^{-t}; \; K(x, t) = xt + 2x^2t^2; \; K(x, t) = x - t.$$

The Fredholm equation with a degenerate kernel can be solved by reducing it to a linear algebraic system of equations.

Consider the Fredholm equation of the second kind (3.17) with a degenerate kernel (3.18). Dividing the integral in (3.17) in accordance with (3.18) into n terms and removing the factor $a_k(x)$ in each of them outside the integral, we write the equation in the form

$$\varphi(x) - \lambda \sum_{k=1}^{n} a_k(x) \int_a^b b_k(t)\varphi(t)\mathrm{d}t = 0. \tag{3.19}$$

Denoting

$$\int_a^b b_k(t)\varphi(t)\mathrm{d}t = c_k, \tag{3.20}$$

we obtain an expression for the unknown function in Eq. (3.19)

$$\varphi(x) = \lambda \sum_{k=1}^{n} c_k a_k(x), \tag{3.21}$$

where c_k is unknown constants.

Substituting expression (3.21) into the integral Eq. (3.17) and taking into account the linear independence of the functions $a_k(x)$, we obtain

$$c_j - \lambda \sum_{k=1}^{n} c_k \int_a^b a_k(t)b_j(t)\mathrm{d}t = 0, \; j = 1, 2 \cdots n.$$

With the notations

$$a_{kj} = \int_a^b a_k(t)b_j(t)\mathrm{d}t, \tag{3.22}$$

we obtain a homogeneous system of algebraic equations

$$c_j - \lambda \sum_{k=1}^{n} c_k a_{kj} = 0, \; j = 1, 2 \cdots n, \tag{3.23}$$

or in expanded form

$$\begin{cases} (1 - \lambda a_{11})c_1 + (-\lambda a_{12})c_2 + \cdots + (-\lambda a_{1n})c_n = 0 \\ (-\lambda a_{21})c_1 + (1 - \lambda a_{22})c_2 + \cdots + (-\lambda a_{2n})c_n = 0 \\ \qquad\qquad \cdots \\ (-\lambda a_{n1})c_1 + (-\lambda a_{n2})c_2 + \cdots + (1 - \lambda a_{nn})c_n = 0 \end{cases}.$$

Let us compose the determinant of this system and equate it to zero

$$\Delta(\lambda) = \begin{vmatrix} 1 - \lambda a_{11} & -\lambda a_{12} & \cdots & -\lambda a_{1n} \\ -\lambda a_{21} & 1 - \lambda a_{22} & \cdots & -\lambda a_{2n} \\ & & & \\ -\lambda a_{n1} & -\lambda a_{n2} & \cdots & 1 - \lambda a_{nn} \end{vmatrix} = 0. \qquad (3.24)$$

The characteristic numbers will be the roots of the algebraic Eq. (3.24).

If Eq. (3.24) has p roots, then the original integral equation has p characteristic numbers, and each characteristic number $\lambda_m, 1 \le m \le p$ corresponds to the nonzero solution $\left(c_1^{(m)}, c_2^{(m)}, \ldots, c_1^{(m)}\right)$. The nonzero solutions of the integral Eq. (3.19) corresponding to these solutions, i.e., eigenfunctions will look like:

$$\varphi_m(x) = \lambda \sum_{k=1}^{n} c_k^{(m)} a_k(x), 1 \le m \le p.$$

Thus, a homogeneous Fredholm integral equation with a degenerate kernel (3.19) has no more than n characteristic numbers and the corresponding eigenfunctions. In this case, the homogeneous Fredholm integral equation may not have characteristic numbers and eigenfunctions at all, or it may not have real characteristic numbers and eigenfunctions.

It is also worth recalling that Volterra's equations do not have characteristic numbers, since all values of the parameter are λ regular. This means that Volterra's equations have no eigenfunctions either.

Let us consider the scheme of the method step by step using examples below.

Example 3.7 Find the characteristic numbers and eigenfunctions of the integral equation:

$$\varphi(x) - \lambda \int_0^\pi \left(\cos 2t \cos^2 x + \cos^3 t \cos 3x\right)\varphi(t)dt = 0.$$

Solution

We will solve the equation in the *Wolfram Mathematica* package. Since the kernel is degenerate, then functions $a_k(x)$ and $b_j(t)$ can be specified as input data separately.

```
In [1]:=Clear[λ,a,b,c,x,t,ϕ]
    ak=Function[{x},{Cos[x]^2,Cos[3x]}];
    bk=Function[{t},{Cos[2t],Cos[t]^3}];
    a=0;
    b=π;
    eqn=ϕ[x]==λ Integrate[ak[x].bk[t] ϕ[t],{t,a,b}]
```

$$\text{Out[6]} := \phi[x] == \lambda \int_0^\pi \left(Cos[2t]Cos[x]^2 + Cos[t]^3 Cos[3x] \right) \phi[t] dt$$

Let us compose the system of Eq. (3.23). To do this, we will create a list of unknown variables c_k and a matrix of coefficients a_{kj}.

```
In [7]:=n=Length@ak[x];
    cList=Subscript[c,#]&/@Range[n]
    coefMatrix=Outer[Integrate[#1*#2,
    {t,a,b}]&,bk[t],ak[t]]
```

$$\text{Out[8]} := \{c_1, c_2\}$$
$$\text{Out[9]} := \left\{ \left\{ \frac{\pi}{4}, 0 \right\}, \left\{ 0, \frac{\pi}{8} \right\} \right\}$$

After that, we can display the system (if required) and find its solutions, which should depend not only on c_k, but also on λ. Therefore, it is more convenient to use the **Reduce** function.

```
In [10]:=system=Thread[(IdentityMatrix
    [n]-λ coefMatrix).cList==0]
```

$$\text{Out[10]} := \left\{ \left(1 - \frac{\pi\lambda}{4} \right) c_1 == 0, \left(1 - \frac{\pi\lambda}{8} \right) c_2 == 0 \right\}$$

```
In [11]:=systemsol={ToRules[Reduce[system,cList]]}
```

$$\text{Out[11]} := \left\{ \left\{ \lambda \to \frac{4}{\pi}, c_2 \to 0 \right\}, \left\{ \lambda \to \frac{8}{\pi}, c_1 \to 0 \right\}, \left\{ c_1 \to 0, c_2 \to 0 \right\} \right\}$$

We see that it is possible to find all solutions of the system, including the one that corresponds to the trivial solution of the original equation.

Now, we will compose the general form of the solution and find the characteristic numbers and the corresponding eigenfunctions. We will discard the trivial solution.

```
In [12]:=gensol=λ cList.ak[x]
```

$\text{Out[12]} := \lambda \left(\text{Cos}[x]^2 c_1 + \text{Cos}[3x] c_2 \right)$

```
In [13]:=eigens=Select[{λ,gensol}/.
    systemsol,!(#[[2]]===0)&];eigens//Column
```

$\text{Out[14]} := \left\{ \frac{4}{\pi}, \frac{4\text{Cos}[x]^2 c_1}{\pi} \right\}$
$\left\{ \frac{8}{\pi}, \frac{8\text{Cos}[3x] c_2}{\pi} \right\}$

Further, we can compose separate lists of characteristic numbers and eigenfunctions. Moreover, since the eigenfunctions are determined up to a constant factor, all constants can be discarded.

```
In [15]:=charValues=eigens[[All,1]]eigenFunctions=
    Select[#//FullSimplify,!FreeQ[#,x]&]&/@eigens[[All,2]]
```

$\text{Out[15]} := \left\{ \frac{4}{\pi}, \frac{8}{\pi} \right\}$
$\text{Out[16]} := \left\{ \text{Cos}[x]^2, \text{Cos}[3x] \right\}$

As a result, we get a list of nontrivial solutions. We can substitute them into the original equation for the verification.

```
In [17]:=sol={λ->charValues[[#]],φ->Function[{x},eigen
    Functions[[#]]//Evaluate]}&/@
    Range[n];
    sol//Column
```

$\text{Out[18]} := \left\{ \lambda \to \frac{4}{\pi}, \phi \to \text{Function}\left[\{x\}, \text{Cos}[x]^2 \right] \right\}$
$\left\{ \lambda \to \frac{8}{\pi}, \phi \to \text{Function}[\{x\}, \text{Cos}[3x]] \right\}$

```
In [19]:=eqn/.sol//FullSimplify
```

$\text{Out[19]} := \{\text{True}, \text{True}\}$

Let us see how the families of eigenfunctions look on the interval $x \in [0, \pi]$ (Figs. 3.6 and 3.7).

```
In [20]:=optplot={AxesStyle->Arrowheads[{0.0,0.025}],
    AxesLabel->{Style[x,Bold,Medium],Style[y,Bold,Medium]}};
    Plot[#eigenFunctions[[1]]//Evaluate,{x,a,b},
    Evaluate@optplot,
    PlotLegends->ToString/@Thread[c1==#]]&@Range[-4,4]
```

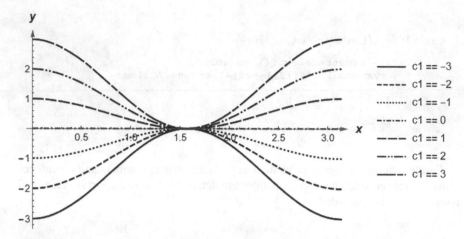

Fig. 3.6 Family of eigenfunctions $\varphi_1(x)$ for various values of an arbitrary constant c_1 (Example 3.7)

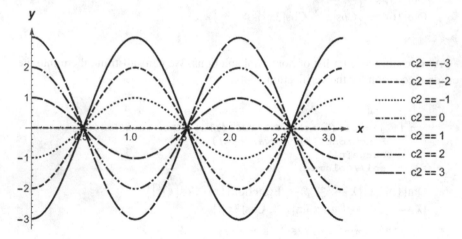

Fig. 3.7 Family of eigenfunctions $\varphi_2(x)$ for various values of an arbitrary constant c_2 (Example 3.7)

```
Plot[# eigenFunctions[[2]]//Evaluate,{x,a,b},
Evaluate@optplot,
PlotLegends->ToString/@Thread[c2==#]]&@Range[-4,4]
```

All the steps can be recorded as a function. In this case, we immediately take into account the cases when equation does not have real characteristic numbers, but may have complex ones.

```
In [23]:=Clear[FredholmEigenFunctions]
         FredholmEigenFunctions::noreal="The integral
equation
         has no real characteristic values.";
         FredholmEigenFunctions::complex="The integral
equation
         has some complex characteristic values:\n`1`.";
         FredholmEigenFunctions[ak_,bk_,{a_,b_},λ_,φ_,x_]:=
         Block[{n,cList,coefMatrix,system,systemsol,gensol,
         eigens,eigensReals,eigensComplexes,charValuesReals,
         charValuesComplexes,eigenFunctionsReals,sols,t,c},
         n=Length@ak[t];
         cList=Subscript[c,#]&/@Range[n];
         coefMatrix=Outer[Integrate[#1*#2,{t,a,b}]&,
         bk[t],ak[t]];
         system=Thread[(IdentityMatrix[n]-λ
coefMatrix).cList==0];
         systemsol={ToRules[Reduce[system,cList]]};
         gensol=λ cList.ak[x];
         eigens=Select[{λ,gensol}/.systemsol,#[[1]]Complexes&];
             eigensReals=Select[eigens,#[[1]]Reals&];
         eigensComplexes=Complement[eigens,eigensReals];
          If[eigensComplexes!={},
         charValuesComplexes=eigensComplexes[[All,1]];
         Message[FredholmEigenFunctions::complex,
         charValuesComplexes]
         ];
             If[eigensReals=={},
         Message[FredholmEigenFunctions::noreal];
         {},
         charValuesReals=eigensReals[[All,1]];
         eigenFunctionsReals=Select[#//FullSimplify,
         !FreeQ[#,x]&]&/@eigensReals[[All,2]];
         {λ->charValuesReals[[#]],
         φ->Function[{t},eigenFunctionsReals[[#]]//
         Evaluate]}&/@Range[Length@eigensReals]/.t->x
         ]
             ]
```

Let us check this function on the current example.

```
In [27]:=FredholmEigenFunctions[ak,bk,{a,b},λ,φ,x]//Column
```

$$\text{Out[27]} := \left\{ \lambda \to \tfrac{4}{\pi}, \phi \to \textbf{Function}\big[\{x\}, \text{Cos}[x]^2\big] \right\}$$
$$\left\{ \lambda \to \tfrac{8}{\pi}, \phi \to \textbf{Function}[\{x\}, \text{Cos}[3x]] \right\}$$

Answers

1. If $\lambda = \frac{4}{\pi}$, eigenfunction is $\varphi(x) = \cos^2 x$.
2. If $\lambda = \frac{8}{\pi}$, eigenfunction is $\varphi(x) = \cos 3x$.

Example 3.8 Find the characteristic numbers and eigenfunctions of the integral equation:

$$\varphi(x) - \lambda \int_0^1 \left(\sqrt{x}t - \sqrt{t}x\right)t\varphi(t)\mathrm{d}t = 0.$$

Solution

Let us set the initial data and use the **FredholmEigenFunctions** function from the previous example.

```
In [28]:=Clear[λ,a,b,c,x,t,φ]
          ak=Function[{x},{Sqrt[x],-x}]];
          bk=Function[{t},{t,Sqrt[t]}];
          a=0;
          b=1;
          eqn=φ[x]==λ Integrate[ak[x].bk[t] φ[t],{t,a,b}]
```

$$\mathrm{Out}[33] := \phi[x] == \lambda \int_0^1 (t\sqrt{x} - \sqrt{t}x)\phi[t]\mathrm{d}t$$

```
In [34]:=FredholmEigenFunctions[ak,bk,{a,b},λ,φ,x]
```

This integral equation has no real characteristic values

```
          FredholmEigenFunctions::complex: The integral
equation has some complex characteristic values:
          {-5 I Sqrt[6],5 I Sqrt[6]}.
          FredholmEigenFunctions::noreal: The integral
equation has no real characteristic values.
```

$$\mathrm{Out}[34] := \{\}$$

Here, we receive a message that this equation does not have real characteristic numbers, but it does have complex ones.

Answer

This equation has no real characteristic numbers and eigenfunctions.

Example 3.9 Find the characteristic numbers and eigenfunctions of the integral equation:

$$\varphi(x) - \lambda \int_0^1 (3x - 2)t\varphi(t)\mathrm{d}t = 0.$$

Solution
Let us set the initial data and use the **FredholmEigenFunctions** function.

```
In [35]:=Clear[λ,a,b,c,x,t,φ]
     ak=Function[{x},{3x,-2}}];
     bk=Function[{t},{t,t}];
     a=0;
     b=1;
     eqn=φ[x]==λ Integrate[ak[x].bk[t] φ[t],{t,a,b}]
```

$$\text{Out[40]} := \phi[x] == \lambda \int_0^1 (-2t + 3tx)\phi[t]\mathrm{d}t$$

```
In [41]:=FredholmEigenFunctions[ak,bk,{a,b},λ,φ,x]//Column
             FredholmEigenFunctions::noreal: The integral
equation has no real characteristic values.
```

$$\text{Out[41]} := \{\}$$

We receive a message in the solution that this equation has no characteristic numbers, while we did not receive a message that there are complex characteristic numbers. This means that this equation has no characteristic numbers at all.

Answer
This equation has no characteristic numbers and eigenfunctions.

3.4 Solution of Fredholm Inhomogeneous Integral Equations with a Degenerate Kernel. Fredholm's Theorems

Consider a Fredholm equation of the second kind with a degenerate kernel

$$\varphi(x) = \lambda \int_a^b \sum_{k=1}^n a_k(x)b_k(t)\varphi(t)\mathrm{d}t + f(x). \tag{3.25}$$

Let us perform the same procedure with the equation as in the previous section. As a result, we find that the solution to Eq. (3.25) can be written as

$$\varphi(x) = \lambda \sum_{t=1}^{n} c_k a_k(x) + f(x), \tag{3.26}$$

where

$$c_k = \int_a^b b_k(t)\varphi(t)\mathrm{d}t$$

is unknown constants that can be found from a system of algebraic equations

$$c_j - \lambda \sum_{k=1}^{n} c_k a_{kj} = f_j, \, j = 1, 2, \ldots, n, \tag{3.27}$$

where

$$a_{kj} = \int_a^b a_k(t)b_j(t)\mathrm{d}t, \, f_j = \int_a^b f(t)b_j(t)\mathrm{d}t. \tag{3.28}$$

Consider the determinant of the system (3.27)

$$\Delta(\lambda) = \begin{vmatrix} 1 - \lambda a_{11} & -\lambda a_{12} & \cdots & -\lambda a_{1n} \\ -\lambda a_{21} & 1 - \lambda a_{22} & \cdots & -\lambda a_{2n} \\ & & & \\ -\lambda a_{n1} & -\lambda a_{n2} & \cdots & 1 - \lambda a_{nn} \end{vmatrix}.$$

1. $\Delta(\lambda) \neq 0$.

In this case, system (3.27) has a unique solution, and the corresponding values of the parameter λ are regular. Moreover, since the matrix of the system depends only on the kernel of the integral equation, a unique solution exists for any free term of the equation.

These results are formulated in the **Fredholm Theorem I**: For a Fredholm integral equation of the second kind to have a unique solution for any free term, it is necessary and sufficient that the corresponding homogeneous equation has only a trivial solution.

2. $\Delta(\lambda) = 0$.

In this case, the corresponding parameter values λ are characteristic. As we remember from the previous paragraph, in this case, each characteristic number λ_m corresponds to a certain number of p_m linearly independent eigenfunctions. Recall that this number p_m is called the rank of the characteristic number λ_m.

Let us formulate a statement concerning the rank of characteristic numbers. For this, we need the concept of associated kernels.

The kernel $K^*(x, t)$ is called adjoint with the kernel $K(x, t)$, if $K^*(x, t) = K(t, x)$. For example, if $K(x, t) = xt^2 - x$, then $K^*(x, t) = tx^2 - t$.

An integral equation with an adjoint kernel is called adjoint with an original equation. Solving mutually adjoint homogeneous integral equations with a degenerate kernel is reduced to solving linear algebraic homogeneous systems of equations with transposed matrices.

Fredholm Theorem II: adjoint kernels have the same characteristic values of the same rank. Otherwise: if λ is the characteristic number of the kernel $K(x, t)$, then the homogeneous integral equation and its adjoint equation have the same finite number of linearly independent eigenfunctions corresponding to λ.

In the case $\Delta(\lambda) = 0$ for inhomogeneous integral Eq. (3.25), **Fredholm Theorem III** is valid: The inhomogeneous integral equation with the characteristic value of the parameter λ is solvable if and only if the free term of the equation is orthogonal to all solutions of the adjoint homogeneous equation.

Orthogonality means the fulfillment of the following condition

$$(f, \psi) = \int_a^b f(x)\psi(x)dx = 0, \tag{3.29}$$

where $f(x)$ is the free term of the Fredholm Eq. (3.25) with the kernel $K(x, t)$; $\psi(t)$ is any solution of the adjoint homogeneous equation for a given characteristic value λ:

$$\psi(x) = \lambda \int_a^b K^*(x, t)\psi(x)dx.$$

Condition (3.29) is sufficient to verify only for linearly independent eigenfunctions of the adjoint equation:

$$(f, \psi_m) = 0, l = 1, 2, \ldots, p \tag{3.30}$$

If conditions (3.30) are satisfied, then the inhomogeneous Fredholm Eq. (3.25) has an infinite set of solutions. They can be found by solving the linear algebraic system (3.27), which has an infinite set of solutions if $\Delta(\lambda) = 0$ and (3.30) are fulfilled. We obtain the general solution of the equation in the form (3.26), where arbitrary constants will enter the expressions for c_k.

If conditions (3.30) are not satisfied, then system (3.27) and the integral Eq. (3.2) have no solution.

Example 3.10 Solve the equation

$$\varphi(x) = 1 + \lambda \int_0^1 (x - t)\varphi(t)dt.$$

Solution

We will solve the equation in *Wolfram Mathematica* using our results from the examples of Sect. 3.3, supplementing the code if needed. Let us set the initial data.

```
In [1]:=Clear[λ,a,b,f,c,x,t,ϕ]
        ak=Function[{x},{x,-1}];
        bk=Function[{t},{1,t}];
        f=Function[{x},1];
        a=0;
        b=1;
        eqn=ϕ[x]==f[x]+λ Integrate[ak[x].bk[t] ϕ[t],{t,a,b}]
```

$$\text{Out}[7] := \phi[x] == 1 + \lambda \int_0^1 (-t + x)\phi[t]dt$$

To construct the system of equations (3.27), we will create a list of unknown variables c_k, a matrix of coefficients a_{kj} and a list of right-hand sides f_k. We also obtain the general form of the solution by using formula (3.26).

```
In [8]:=n=Length@ak[x];
        cList=Subscript[c,#]&/@Range[n];
        coefMatrix=Outer[Integrate[#1*#2,{t,a,b}]&,bk[t],ak[t]]
        fList=Integrate[f[t]*#,{t,a,b}]&/@bk[t]
```

$$\text{Out}[9] := \{c_1, c_2\}$$
$$\text{Out}[10] := \left\{\left\{\tfrac{1}{2}, -1\right\}, \left\{\tfrac{1}{3}, -\tfrac{1}{2}\right\}\right\}$$
$$\text{Out}[11] := \left\{1, \tfrac{1}{2}\right\}$$

```
In [12]:=gensol=f[x]+λ cList.ak[x]
```

$$\text{Out}[12] := 1 + \lambda(xc_1 - c_2)$$

Let us compose and display the system. Then, we find the characteristic numbers, equating the determinant of the system to zero.

```
In [13]:=system=Thread[(IdentityMatrix[n]-
          λ coefMatrix).cList==fList];system//Column
```

Out[14] := $\left(1 - \frac{\lambda}{2}\right)c_1 + \lambda c_2 == 1$
$-\frac{\lambda c_1}{3} + \left(1 + \frac{\lambda}{2}\right)c_2 == \frac{1}{2}$

In   [15]:=Δλroots=Solve[Det[(IdentityMatrix[n]-λ coefMatrix)]==0,λ]

Out[15] := $\left\{\left\{\lambda \rightarrow -2i\sqrt{3}\right\}, \left\{\lambda \rightarrow 2i\sqrt{3}\right\}\right\}$

This means that all real values λ will be regular, i.e., the system of algebraic equations and the original integral equation have a unique solution for any real value λ. Let us solve the system and find a solution of the original equation.

In [16]:=systemsol=Solve[system,cList][[1]]//FullSimplify

Out[16] := $\left\{c_1 \rightarrow \frac{12}{12+\lambda^2}, c_2 \rightarrow \frac{6+\lambda}{12+\lambda^2}\right\}$

In [17]:= sol=φ->Function[{x},gensol/.systemsol// FullSimplify//Evaluate]

Out[17] := $\phi \rightarrow$ **Function**$\left[\{x\}, \frac{6(2+(-1+2x)\lambda)}{12+\lambda^2}\right]$

Let us check the solution by substituting it into the original equation.

In [18]:=eqn/.sol//FullSimplify
Out[18]:=True

Answer

$$\varphi(x) = \frac{6(2 + 2\lambda x - \lambda)}{12 + \lambda^2}, \quad \text{for any real} \quad \lambda.$$

Example 3.11 Check the validity of Fredholm Theorem II for the kernel $K(x, t) = xt^2 - x^2$ if $x, t \in [-1; 1]$.

Solution
To check the validity of Fredholm Theorem II, we need to find the characteristic numbers and eigenfunctions for a given kernel and its adjoint one. We can use the **FredholmEigenFunctions** we defined earlier.

Let us start the cell with the function definition and set the initial data.

```
In [5]:=Clear[λ,a,b,c,x,t,φ]
    ak=Function[{x},{x,-x^2}];
    bk=Function[{t},{t^2,1}];
    a=-1;
    b=1;
```

Let us run **FredholmEigenFunctions** to find the characteristic numbers and eigenfunctions of an original kernel. To find them for adjoint kernel, we need to interchange the variables *ak* and *bk* in function call.

```
In [10]:=FredholmEigenFunctions[ak,bk,{a,b},λ,φ,x]
```
$$\text{Out}[10] := \left\{ \left\{ \lambda \to -\tfrac{3}{2}, \phi \to \textbf{Function}[\{x\}, x(-3+5x)] \right\} \right\}$$
```
In [11]:=FredholmEigenFunctions[bk,ak,{a,b},λ,φ,x]
```
$$\text{Out}[11] := \left\{ \left\{ \lambda \to -\tfrac{3}{2}, \phi \to \textbf{Function}[\{x\}, 1] \right\} \right\}$$

Answer

Thus, we get that the original kernel and its adjoint kernel have the same characteristic numbers of the same rank. The validity of Fredholm Theorem II is verified using the following example.

Example 3.12 Explore and solve the equation for different values of the parameter λ.

$$\varphi(x) = \lambda \int_0^{\pi} \cos(x+t)\varphi(t)\mathrm{d}t + \cos 3x.$$

Solution

The kernel of the equation is degenerate:

$$\cos(x+t) = \cos x \cos t - \sin x \sin t.$$

The input data in this case looks like this:

```
In [1]:=Clear[λ,a,b,f,c,x,t,φ]
    ak=Function[{x},{Cos[x],-Sin[x]}];
    bk=Function[{t},{Cos[t],Sin[t]}];
    f=Function[{x},Cos[3x]];
    a=0;
```

```
b=π;
eqn=φ[x]==f[x]+λ Integrate[ak[x].bk[t] φ[t],{t,a,b}]
```

$$Out[7] := \phi[x] == Cos[3x] + \lambda \int_0^\pi (Cos[t]Cos[x] - Sin[t]Sin[x])\phi[t]dt$$

Set the necessary variables for constructing a system of algebraic equations. We will also set the general form of the solution.

```
In [8]:=n=Length@ak[x];
        cList=Subscript[c,#]&/@Range[n];
        coefMatrix=Outer[Integrate[#1*#2,{t,a,b}]&,bk[t],ak[t]];
        fList=Integrate[f[t]*#,{t,a,b}]&/@bk[t];
        gensol=f[x]+λ cList.ak[x]
```

$$Out[12] := Cos[3x] + \lambda(Cos[x]c_1 - Sin[x]c_2)$$

We build a system and find its solutions.

```
In [13]:=system=Thread[(IdentityMatrix[n]-
          λ coefMatrix).cList==fList];system//Column
```

$$Out[14] := \begin{matrix} \left(1 - \frac{\pi\lambda}{2}\right)c_1 == 0 \\ \left(1 + \frac{\pi\lambda}{2}\right)c_2 == 0 \end{matrix}$$

```
In[15]:=systemsol={ToRules[Reduce[system,cList]]}
```

$$Out[15] := \left\{\left\{\lambda \to -\tfrac{2}{\pi}, c_1 \to 0\right\}, \left\{\lambda \to \tfrac{2}{\pi}, c_2 \to 0\right\}, \{c_1 \to 0, c_2 \to 0\}\right\}$$

It can be seen that one solution corresponds to regular values of the parameter λ, and the other two correspond to characteristic values. Let us try to obtain it.

```
In [16]:=allsols={λ,gensol}/.systemsol;
         eigensols=Select[{λ,gensol}/.systemsol,#[[1]]Reals&]
         regularsols=Complement[allsols,eigensols]
```

$$Out[17] := \left\{\left\{-\tfrac{2}{\pi}, Cos[3x] + \tfrac{2Sin[x]c_2}{\pi}\right\}, \left\{\tfrac{2}{\pi}, Cos[3x] + \tfrac{2Cos[x]c_1}{\pi}\right\}\right\}$$
$$Out[18] := \{\{\lambda, Cos[3x]\}\}$$

Now, all that remains is to formalize the result. We will also do the usual verification of solutions by setting them into the original equation.

```
In [19]:=charValues=eigensols[[All,1]];
        eigenFunctions=eigensols[[All,2]];
        regsol={ϕ->Function[{x},#]}&/@regularsols[[All,2]]
        irregsol={λ->charValues[[#]],
        ϕ->Function[{x},eigenFunctions[[#]]//Evaluate]}&/@
        Range[Length@charValues]
```

Out[21] := {{$\phi \to$ **Function**[{x}, **Cos**[$3x$]]}}

Out[22] := {{$\lambda \to -\frac{2}{\pi}, \phi \to$ **Function**[{x}, **Cos**[$3x$]$+\frac{2\text{Sin}[x]c_2}{\pi}$]},

{$\lambda \to frac2\pi, \phi \to$ **Function**[{x}, **Cos**[$3x$]$+\frac{2\text{Cos}[x]c_1}{\pi}$]}}}

```
In [23]:=eqn/.regsol//FullSimplifyeqn/.irregsol//FullSimplify
```

Out[23] := {**True**}

Out[24] := {**True**, **True**}

This gives us three different solutions.

1. The unique solution for regular values $\lambda \neq \pm\frac{2}{\pi}$:

$$\varphi(x) = \cos 3x.$$

2. An infinite number of solutions for the characteristic value $\lambda = -\frac{2}{\pi}$:

$$\varphi(x) = \frac{2c_2 \sin x}{\pi} + \cos 3x = \tilde{c} \sin x + \cos 3x, \forall \tilde{c}.$$

3. An infinite number of solutions for the characteristic value $\lambda = \frac{2}{\pi}$:

$$\varphi(x) = \frac{2c_1 \cos x}{\pi} + \cos 3x = \tilde{c} \cos x + \cos 3x, \forall \tilde{c}.$$

The graphs for all solutions are displayed in Fig. 3.8 Fig. 3.9Fig. 3.10.

```
In [25]:=optplot={AxesStyle->Arrowheads[{0.0,0.025}],
        AxesLabel->{Style[x,Bold,Medium],
        Style[y,Bold,Medium]}};
        Plot[ϕ[x]/.regsol//Evaluate,{x,a,b},
        Plot[ϕ[x]/.regsol//Evaluate,{x,a,b},Evaluate@optplot]
        Evaluate@optplot]
        Evaluate@optplot,
```

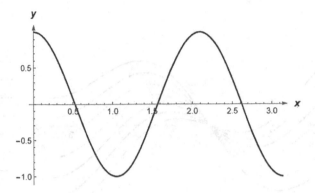

Fig. 3.8 Solution of an integral equation with regular values λ (Example 3.12)

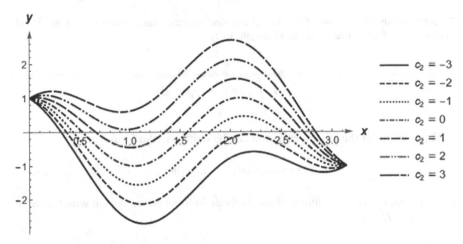

Fig. 3.9 Family of solutions of an integral equation with the characteristic value $\lambda = -\frac{2}{\pi}$ and various values of an arbitrary constant (Example 3.12)

```
PlotLegends->Thread[c2==#]]&@Range[-4,4]
Plot[φ[x]/.irregsol[[2]]/.c1->#//Evaluate,{x,a,b},
Evaluate@optplot,
PlotLegends->Thread[c1==#]]&@Range[-4,4]
```

The cases of the characteristic values of the parameter $\lambda = \pm\frac{2}{\pi}$ can also be investigated by using Fredholm Theorem III. For the original kernel, $K(x, t) = \cos(x + t)$ and adjoint kernel $K^*(x, t) = \cos(t + x) = K(x, t)$. Consequently, the adjoint homogeneous equation has the form

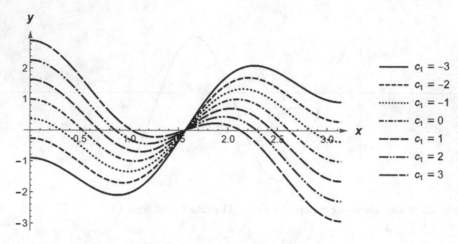

Fig. 3.10 Family of solutions of an integral equation with the characteristic value $\lambda = \frac{2}{\pi}$ and various values of an arbitrary constant (Example 3.12)

$$\psi(x) = \lambda \int_0^\pi \cos(x+t)\psi(x)\mathrm{d}x.$$

The solution of the adjoint equation can be written in the form

$$\psi(x) = \lambda c_1 \cos x - \lambda c_2 \sin x, \tag{3.31}$$

where c_1 and c_2 are the solution of an algebraic homogeneous system with the same determinant $D(\lambda)$:

$$\begin{cases} c_1\left(1 - \frac{\lambda\pi}{2}\right) = 0 \\ c_2\left(1 + \frac{\lambda\pi}{2}\right) = 0 \end{cases}. \tag{3.32}$$

For the characteristic value $\lambda = \frac{2}{\pi}$, system (3.32) takes the form

$$\begin{cases} c_1 \cdot 0 = 0; \\ c_2 \cdot 2 = 0. \end{cases}$$

So, we get that c_1 is arbitrary constant, $c_2 = 0$. Then, in accordance to (3.37) if $\lambda = \frac{2}{\pi}$,

$$\psi(x) = \frac{2}{\pi}c_1 \cos x.$$

Here, the number $\lambda = \frac{2}{\pi}$ corresponds to one independent eigenfunction

$$\psi_1(x) = \cos x.$$

By Fredholm Theorem III, the inhomogeneous equation for $\lambda = \frac{2}{\pi}$ is solvable if and only if $(f, \psi_1) = 0$. Since

$$(f, \psi_1) = \int\limits_0^\pi \cos 3x \cos x \, dx = 0,$$

and this inhomogeneous equation for $\lambda = \frac{2}{\pi}$ has an infinite number of solutions. For the characteristic value $\lambda = -\frac{2}{\pi}$, system (3.32) takes the form

$$\begin{cases} c_1 \cdot 2 = 0; \\ c_2 \cdot 0 = 0. \end{cases}$$

Hence, $c_1 = 0, c_2$ is arbitrary constant. Then, in accordance with (3.37), $\lambda = -\frac{2}{\pi}$ corresponds one independent eigenfunction $\psi_2(x) = \sin x$. To investigate the solvability of an inhomogeneous equation for $\lambda = -\frac{2}{\pi}$, it is necessary to calculate (f, ψ_2). Since

$$(f, \psi_2) = \int\limits_0^\pi \cos 3x \sin x \, dx = 0,$$

by Fredholm Theorem III, the inhomogeneous equation for $\lambda = -\frac{2}{\pi}$ also has an infinite set of solutions.

Answers

1. $\varphi(x) = \cos 3x$ if $\lambda \neq \pm\dfrac{2}{\pi}$.

2. $\varphi(x) = \tilde{c} \sin x + \cos 3x, \forall \tilde{c}, if \lambda = -\dfrac{2}{\pi}$.

3. $\varphi(x) = \tilde{c} \cos x + \cos 3x, \forall \tilde{c}, if \lambda = \dfrac{2}{\pi}$.

Example 3.13 Solve the integral equation:

$$\varphi(x) - \lambda \int\limits_0^1 \sin \log(x) \varphi(t) dt = 2x.$$

Solution

This kernel can be considered degenerate and the functions $b_1(t) = 1$ and $a_1(x) = \sin\log(x)$. Then, the data input will look like this:

```
In [29]:=Clear[λ,a,b,f,c,x,t,φ]
         ak=Function[{x},{Sin[Log[x]]}];
         bk=Function[{t},{1}];
         f=Function[{x},2x];
         a=0;
         b=1;
         eqn=φ[x]==f[x]+λ Integrate[ak[x].bk[t] φ[t],{t,a,b}]
```

$$\text{Out[35]} := \phi[x] == 2x + \lambda \int_0^1 \mathbf{Sin}\big[\mathbf{Log}[x]\big]\phi[t]\mathrm{d}t$$

Set the necessary variables for constructing a system of algebraic equations. We will also set the general form of the solution.

```
In [36]:=n=Length@ak[x];
         cList=Subscript[c,#]&/@Range[n];
         coefMatrix=Outer[Integrate[#1*#2,{t,a,b}]&,bk[t],ak[t]];
         fList=Integrate[f[t]*#,{t,a,b}]&/@bk[t];
         gensol=f[x]+λ cList.ak[x]
```

$$\text{Out[40]} := 2x + \lambda \mathbf{Sin}\big[\mathbf{Log}[x]\big]c_1$$

Construct the system and find the characteristic values of the parameter λ.

```
In [41]:=system=Thread[(IdentityMatrix[n]-
         λ coefMatrix).cList==fList];system//Column
```

$$\text{Out[42]} := \big(1 + \tfrac{\lambda}{2}\big)c_1 == 1$$

```
In [43]:=Solve[Det[IdentityMatrix[n]-λ coefMatrix]==0,λ]
```

$$\text{Out[43]} := \{\{\lambda \to -2\}\}$$

It turns out that this equation has only one characteristic value $\lambda = 2$. Let us use the **Reduce** function to look at possible solutions of the algebraic system.

```
In [44]:=Reduce[system,cList]
```

```
Out[44] := 2 + λ ≠ 0&&c₁ == 2/(2+λ)
```

The condition $2 + \lambda \neq 0$ means that the system is unsolvable for $\lambda = -2$. That is, an inhomogeneous equation has a solution only for regular λ. Let us build this solution.

```
In [45]:=systemsol={ToRules[Reduce[system,cList]]}
```

$$Out[45] := \left\{\left\{c_1 \to \frac{2}{2+\lambda}\right\}\right\}$$

```
In [46]:=allsols={λ,gensol}/.systemsol;
    eigensols=Select[{λ,gensol}/.systemsol,#[[1]]Reals&];
    regularsols=Complement[allsols,eigensols];
In[49]:= regsol=ϕ->Function[{x},#]&/@regularsols[[All,2]]
    irregsol={λ->charValues[[#]],
    ϕ->Function[{x},eigenFunctions[[#]]//Evaluate]}&/@
    Range[Length@charValues]
```

$$Out[49] := \left\{\phi \to \mathbf{Function}\left[\{x\}, 2x + \frac{2\lambda \operatorname{Sin}\left[\operatorname{Log}[x]\right]}{2+\lambda}\right]\right\}$$

```
In [50]:=eqn/.regsol//FullSimplify
```

Out[50] := **True**

Thus, this equation has a unique solution for regular values $\lambda \neq -2$:

$$\varphi(x) = \cos 3x$$

and is unsolvable for $\lambda = -2$.

Now, we can plot the graphs of the family of solutions for regular values λ (Fig. 3.11).

```
In [51]:=Plot[ϕ[x]/.regsol/.λ->#//Evaluate,{x,a,b},
    Evaluate@optplot,
    PlotLegends->Thread[λ==#]]&@Complement[Range[-5,3],{-2}]
```

Answers

1. $\varphi(x) = \cos 3x, if \lambda \neq -2$.

2. not solvable if $\lambda = -2$.

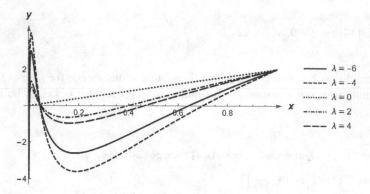

Fig. 3.11 Family of solutions of an integral equation for various regular values λ (Example 3.13)

Example 3.14 Solve the integral equation:

$$\varphi(x) - \lambda \int\limits_{-\pi}^{\pi} \left(x\cos t + t^2 \sin x + \cos x \sin t\right)\varphi(t)\mathrm{d}t = x$$

Solution
Set the initial data.

```
In [52]:=Clear[λ,a,b,f,c,x,t,φ]
        ak=Function[{x},{x,Sin[x],Cos[x]}];
        bk=Function[{t},{Cos[t],t^2,Sin[t]}];
        f=Function[{x},x];
        a=-π;
        b=π;
        eqn=φ[x]==f[x]+λ Integrate[ak[x].bk[t] φ[t],{t,a,b}]
```

$$\mathrm{Out[58]} := \phi[x] == 2x + \lambda \int\limits_{0}^{1} \mathrm{Sin}\big[\mathrm{Log}[x]\big]\phi[t]\mathrm{d}t$$

Set the necessary variables for constructing a system of algebraic equations. We will also set the general form of the solution.

```
In [59]:=n=Length@ak[x];
        cList=Subscript[c,#]&/@Range[n];
        coefMatrix=Outer[Integrate[#1*#2,{t,a,b}]&,bk[t],ak[t]];
        fList=Integrate[f[t]*#,{t,a,b}]&/@bk[t];
        gensol=f[x]+λ cList.ak[x]
```

$$\text{Out[63]} := x + \lambda(x c_1 + \text{Sin}[x]c_2 + \text{Cos}[x]c_3)$$

Construct the system and find the characteristic values of the parameter λ.

```
In [64]:=system=Thread[(IdentityMatrix[n]-
        λ coefMatrix).cList==fList];system//Column
```

$$\text{Out[65]} := \begin{array}{c} c_1 - \pi\lambda c_3 == 0 \\ c_2 + 4\pi\lambda c_3 == 0 \\ -2\pi\lambda c_1 - \pi\lambda c_2 + c_3 == 2\pi \end{array}$$

```
In [66]:=Solve[Det[IdentityMatrix[n]-λ coefMatrix]==0,λ]
```

$$\text{Out[66]} := \left\{ \left\{ \lambda \to -\frac{i}{\sqrt{2\pi}} \right\}, \left\{ \lambda \to \frac{i}{\sqrt{2\pi}} \right\} \right\}$$

We meet a situation of complex roots. In this case, the given integral equation does not have real characteristic numbers and eigenfunctions. This means that all real values of the parameter λ will be regular. Let us use the **Reduce** function to look at possible solutions to the algebraic system.

```
In [67]:=Reduce[system,cList]
```

$$\text{Out[67]} := 1 + 2\pi^2\lambda^2 \neq 0 \&\& c_1 == \frac{2\pi^2\lambda}{1+2\pi^2\lambda^2} \&\& c_2 == -4c_1 \&\& c_3 == -2\pi(-1 + \lambda c_1)$$

This inhomogeneous equation is unsolvable for $\lambda = \pm\frac{i}{\sqrt{2\pi}}$, but we are not interested in these values. Let us construct a solution for regular λ.

```
In [68]:=systemsol={ToRules[Reduce[system,cList]]}
```

$$\text{Out[68]} := \left\{ \left\{ c_1 \to \frac{2\pi^2\lambda}{1+2\pi^2\lambda^2}, c_2 \to -4c_1, c_3 \to -2\pi(-1 + \lambda c_1) \right\} \right\}$$

```
In [69]:=allsols={λ,gensol}//.systemsol;
        eigensols=Select[{λ,gensol}//.systemsol,#[[1]]Reals&];
        regularsols=Complement[allsols,eigensols]//Together;
In[72]:=regsol=ϕ->Function[{x},#]&/@regularsols[[All,2]]
        irregsol={λ->charValues[[#]],
        ϕ->Function[{x},eigenFunctions[[#]]//Evaluate]}&/@
        Range[Length@charValues]
```

$$\text{Out[72]} := \left\{ \phi \to \text{Function}\left[\{x\}, \frac{x+4\pi^2 x\lambda^2 + 2\pi\lambda\text{Cos}[x] - 8\pi^2\lambda^2\text{Sin}[x]}{1+2\pi^2\lambda^2} \right] \right\}$$

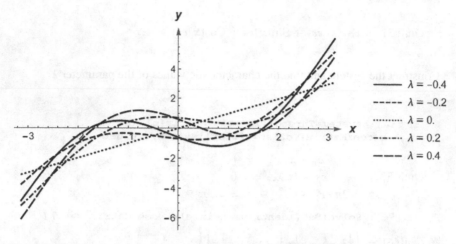

Fig. 3.12 Family of solutions of an integral equation for various regular values λ (Example 3.14)

```
In [73]:=eqn/.regsol//FullSimplify
Out[73] := True
```

Thus, this equation has a unique solution for any real values of the parameter λ:

$$\varphi(x) = \frac{x + 2\pi\lambda\cos x + 4\pi^2\lambda^2 x - 8\pi^2\lambda^2 \sin x}{1 + 2\pi^2\lambda^2}.$$

Figure 3.12 shows the graph of the family of solutions for regular values λ.

```
In [74]:=Plot[ϕ[x]/.regsol/.λ->#//Evaluate,{x,a,b},
          Evaluate@optplot,
          PlotLegends->Thread[λ==#]]&@Range[-0.4,0.4,0.2]
```

Answer

$$\varphi(x) = \frac{x + 2\pi\lambda\cos x + 4\pi^2\lambda^2 x - 8\pi^2\lambda^2 \sin x}{1 + 2\pi^2\lambda^2}, \forall\lambda \in \mathbf{R}.$$

Example 3.15 For which values of α and β, the equation is solvable

$$\varphi(x) = \lambda \int\limits_0^1 xt^2 \varphi(t)\mathrm{d}t + \alpha x + \beta ?$$

Solution

Solvability can be checked in accordance with Fredholm Theorem III. For the characteristic value of the parameter λ, equation is solvable if the free term of the equation is orthogonal to all solutions of the adjoint homogeneous equation, i.e.,

$$(f, \psi) = \int\limits_a^b f(x)\psi(x)\mathrm{d}x = 0,$$

where $f(x)$ is the free term and $\psi(x)$ is the solution of the adjoint homogeneous equation.

We will solve this example in *Wolfram Mathematica*. Let us use the **FredholmEigenFunctions** function defined in Example 3.7. Let us set the initial data.

```
In [5]:=Clear[λ,a,b,f,c,x,t,ϕ,α,β]
        ak=Function[{x},{x}];
        bk=Function[{t},{t^2}];
        f=Function[{x},α x+β];
        a=0;
        b=1;
```

Now, let us call the **FredholmEigenFunctions** function to find the eigenfunctions of the adjoint homogeneous equation.

```
In[11]:=eigens=FredholmEigenFunctions[bk,ak,{a,b},λ,ϕ,x]
```
$$\text{Out[11]} := \left\{\left\{\lambda \to 4, \phi \to \mathbf{Function}\big[\{x\}, x^2\big]\right\}\right\}$$

Thus, the associated equation has a unique eigenfunction $\psi(x) = x^2$, corresponding to the characteristic number $\lambda = 4$. Let us calculate (f, ψ):

```
In [12]:=Integrate[f[x]ϕ[x]/.eigens,{x,a,b}]
```
$$\text{Out[12]} := \left\{\frac{\alpha}{4} + \frac{\beta}{3}\right\}$$

Answers

The original equation is solvable:

1. if $\lambda \neq 4$ for any α and β (the unique solution);
2. if $\lambda = 4$ for α and β, satisfying the condition $\frac{\alpha}{4} + \frac{\beta}{3} = 0$ (an infinite number of solutions).

Example 3.17 Find all solutions of the integral equation for all λ and for all values of the parameters α and β included in the free term of the equation

$$\varphi(x) = \lambda \int\limits_{-1}^{1} (xt + x^2 t^2) \varphi(t) dt + \alpha x + \beta.$$

Solution

Set the initial data.

```
In [13]:=Clear[λ,a,b,f,c,x,t,ϕ,α,β]
        ak=Function[{x},{x,x^2}];
        bk=Function[{t},{t,t^2}];
        f=Function[{x},α x+β];
        a=-1;
        b=1;
        eqn=ϕ[x]==f[x]+λ Integrate[ak[x].bk[t]ϕ[t],{t,a,b}]
```

$$\text{Out[19]} := \phi[x] == x\alpha + \beta + \lambda \int\limits_{-1}^{1} \left(\frac{tx}{2} + \frac{t^2 x^2}{2} \right) \phi[t] dt$$

Set the necessary variables for constructing a system of algebraic equations. We will also set the general form of the solution.

```
In [20]:=n=Length@ak[x];
        cList=Subscript[c,#]&/@Range[n];
        coefMatrix=Outer[Integrate[#1*#2,{t,a,b}]&,bk[t],ak[t]];
        fList=Integrate[f[t]*#,{t,a,b}]&/@bk[t];
        gensol=f[x]+λ cList.ak[x]
```

$$\text{Out[24]} := x\alpha + \beta + \lambda \left(\frac{x c_1}{2} + \frac{x^2 c_2}{2} \right)$$

We construct the system and find the characteristic values of the parameter λ.

```
In [25]:=system=Thread[(IdentityMatrix[n]-
        λ coefMatrix).cList==fList];system//Column
```

$$Out[26] := \begin{matrix} \left(1 - \frac{\lambda}{3}\right)c_1 == \frac{2\alpha}{3} \\ \left(1 - \frac{\lambda}{5}\right)c_2 == \frac{2\beta}{3} \end{matrix}$$

```
In [27]:=Solve[Det[IdentityMatrix[n]-λ coefMatrix]==0,λ]
```

$$Out[27] := \{\{\lambda \to 3\}, \{\lambda \to 5\}\}$$

This means that this equation has two characteristic values of the parameter: $\lambda = 3$ and $\lambda = 5$. Let us use the **Reduce** function to get possible solutions of the algebraic system.

```
In [28]:=Reduce[system,cList]
```

$$Out[28] :=$$

$$\left(\lambda == 3 \&\& \alpha == 0 \&\& c_2 == \frac{5\beta}{3}\right)$$

$$||(\lambda == 5 \&\& \beta == 0 \&\& c_1 == -\alpha)||$$

$$\left(-3+\lambda \neq 0 \&\& c_1 == -\frac{2\alpha}{-3+\lambda} \&\& \quad 5 \mid \lambda \neq 0 \&\& c_2 == \frac{10\beta}{3(-5+\lambda)}\right)$$

We see that the system has three solutions, two of which correspond to the characteristic values of the parameter λ and one to regular. Let us construct a solution for regular and characteristic values of λ.

```
In [29]:=systemsol={ToRules[Reduce[system,cList]]}
```

$$Out[29] := \left\{\left\{\lambda \to 3, \alpha \to 0, c_2 \to \frac{5\beta}{3}\right\}, \{\lambda \to 5, \beta \to 0, c_1 \to -\alpha\},\right.$$

$$\left.\left\{c_1 \to -\frac{2\alpha}{-3+\lambda}, c_2 \to -\frac{10\beta}{3(-5+\lambda)}\right\}\right\}$$

```
In [30]:=allsols={λ,gensol}//.systemsol//FullSimplify;
        eigensols=Take[allsols,2];
        regularsols=Complement[allsols,eigensols];
In[33]:=charValues=eigensols[[All,1]];
        eigenFunctions=eigensols[[All,2]];
        regsol=ϕ->Function[{x},#]&/@regularsols[[All,2]]
        irregsol={λ->charValues[[#]],
         ϕ->Function[{x},eigenFunctions[[#]]//Evaluate]}&/@
        Range[Length@charValues]
```

$$\text{Out[35]} := \left\{\left\{\phi \to \textbf{Function}\left[\{x\}, \beta - \frac{3x\alpha}{-3+\lambda} - \frac{5x^2\beta\lambda}{3(-5+\lambda)}\right]\right\}\right\}$$

$$\text{Out[36]} := \left\{\left\{\lambda \to 3, \phi \to \textbf{Function}\left[\{x\}, \beta + \frac{5x^2\beta}{2} + \frac{3xc_1}{2}\right]\right\}\right.$$
$$\left.\left\{\lambda \to 5, \phi \to \textbf{Function}\left[\{x\}, \tfrac{1}{2}x(-3\alpha + 5xc_2)\right]\right\}\right\}$$

Thus, we have three cases:

1. $\lambda \neq 3$ and $\lambda \neq 5$. In this case, the solution is unique and has the form:

$$\varphi(x) = \beta - \frac{3x\alpha}{\lambda - 3} - \frac{5x^2\beta\lambda}{3(\lambda - 5)}.$$

2. $\lambda = 3$. In this case, the equation is solvable only for $\alpha = 0$ (see the values of the variable *systemsol*). The solution is determined up to a constant c_1 and has the form:

$$\varphi(x) = \beta + \frac{5x^2\beta}{2} + \frac{3xc_1}{2}.$$

3. $\lambda = 5$. In this case, the equation is solvable only for $\beta = 0$ (see the values of the variable *systemsol*). The solution is determined up to a constant c_2 and has the form:

$$\varphi(x) = \frac{1}{2}x(5xc_2 - 3\alpha).$$

Check the solutions by using substitution.

```
In [37]:=eqn/.regsol//eqn/.irregsol//FullSimplifyFullSimplify
```
$$\text{Out[37]} := \{\textbf{True}\}$$
$$\text{Out[38]} := \{x\alpha == 0, \beta == 0\}$$

In the last output, we must take into account that it refers to cases 2 and 3, respectively, so this is actually also a true equality.

The solvability of the equations in cases 2 and 3 can also be checked using Fredholm Theorem III. Let us use the **FredholmEigenFunctions** function to find the eigenfunctions of the adjoint homogeneous equation.

```
In [39]:=eigens=FredholmEigenFunctions[bk,ak,{a,b},λ,φ,x]
```
$$\text{Out[39]} := \left\{\{\lambda \to 3, \phi \to \textbf{Function}[\{x\}, x]\}, \{\lambda \to 5, \phi \to \textbf{Function}[\{x\}, x^2]\}\right\}$$

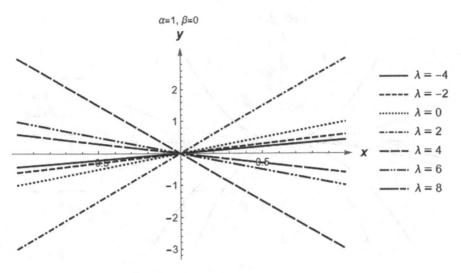

Fig. 3.13 Graph of solutions for regular λ with $\alpha = 1$, $\beta = 0$ (Example 3.16)

Thus, the associated equation has two eigenfunctions: $\psi_1(x) = x$, corresponding to the characteristic number $\lambda = 3$ and $\psi_2(x) = x^2$, corresponding to the characteristic number $\lambda = 5$. Let us calculate (f, ψ):

```
In [40]:=Thread[Integrate[f[x]ϕ[x]/.eigens,{x,a,b}]==0]
```
$$\text{Out[40]} := \left\{ \tfrac{2\alpha}{3} == 0, \tfrac{2\beta}{3} == 0 \right\}$$

This inhomogeneous equation is solvable for $\lambda = 3$, if $\alpha = 0$, and for $\lambda = 5$, if $\beta = 0$.

Figures 3.13 and 3.14 show the graphs corresponding different regular values of λ and different values of α and β.

```
In [41]:=optplot={AxesStyle->Arrowheads[{0.0,0.025}],
         AxesLabel->{Style[x,Bold,Medium],Style[y,Bold,Medium]}};
         Plot[ϕ[x]/.regsol/.{α->1,β->0,λ->#}//Evaluate,{x,a,b},
         PlotLabel->"α=1, β=0",Evaluate@optplot,
         PlotLegends->Thread[λ==#]]&@Range[-4,8,2]
         Plot[ϕ[x]/.regsol/.{α->0,β->1,λ->#}//Evaluate,{x,a,b},
         PlotLabel->"α=0, β=1",Evaluate@optplot,
         PlotLegends->Thread[λ==#]]&@Range[-4,8,2]
```

Figures 3.15 and 3.16 show the graphs for characteristic values of λ.

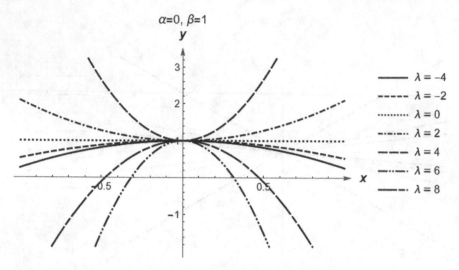

Fig. 3.14 Graph of solutions for regular λ with $\alpha = 0$, $\beta = 1$ (Example 3.16)

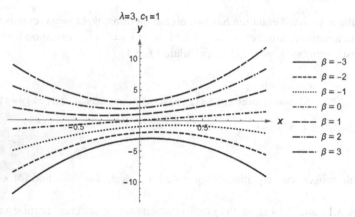

Fig. 3.15 Graph of solutions for characteristic value $\lambda = 3$ with different β (Example 3.16)

```
In[43]:=Plot[φ[x]/.irregsol[[1]]]/.{c1->1,β->#}//Evaluate,{x,a,b},
        PlotLabel->"λ=3, c1=1",Evaluate@optplot,
        PlotLegends->Thread[β==#]]&@Range[-4,4,1]
        Plot[φ[x]/.irregsol[[2]]]/.{c2->1,α->#}//Evaluate,{x,a,b},
        PlotLabel->"λ=5, c2=1",Evaluate@optplot,
        PlotLegends->Thread[α==#]]&@Range[-4,4,1]
```

Answers

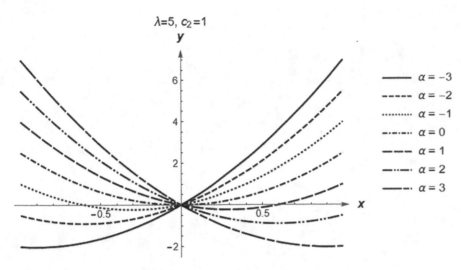

$\lambda=5,\ c_2=1$

Fig. 3.16 Graph of solutions for characteristic value $\lambda = 5$ with different α (Example 3.16)

1. if $\lambda \neq 3$ and $\lambda \neq 5$, $\varphi(x) = \beta - \frac{3x\alpha}{\lambda-3} - \frac{5x^2\beta\lambda}{3(\lambda-5)}$;
2. if $\lambda = 3$ and $\alpha = 0$, $\varphi(x) = \beta + \frac{5x^2\beta}{2} + \frac{3xc_1}{2}$;
3. if $\lambda = 5$ and $\beta = 0$ $\varphi(x) = \frac{1}{2}x(5xc_2 - 3\alpha)$.

References

S. G Mikhlin, *Linear Integral Equations*, Courier Dover Publications, 2020

S. G Mikhlin, *Integral Equations and their Applications to Certain Problems in Mechanics, Mathematical Physics and Technology*, (Pergamon Press, 1964)

I. G Petrovskii, *Lectures on the Theory of Integral Equations*, (Graylock Press, Rochester, 1957)

S.G Mikhlin, On the convergence of Fredholm series, DAN SSSR 42, no. 9, 387–390, 1944

D. Porter, D.S.G. Stirling, *Integral Equations: A Practical Treatment, from Spectral Theory to Applications*. (- Cambridge – New York: Cambridge Univ. Press, 1990)

R.P Kanwal, *Linear Integral Equations*. (– Boston: Birkhäuser, 1996)

J. Kondo, *Integral Equations*. (– Oxford: Clarendon Press, 1991)

A.D Polyanin, A.V Manzhirov, *Handbook of Integral Equations*. (– Boca Raton – New York: CRC Press, 1998)

M. LKrasnov, A. I Kiselev, and G. I. Makarenko, *Problems and Exercises in Integral Equations*, (Mir Publ., Moscow, 1971)

Wolfram Mathematica, http://www.wolfram.com/mathematica/

Chapter 4
Symmetric Integral Equations

4.1 Construction of an Orthonormal System of Eigenfunctions of a Symmetric Kernel

The kernel of an integral equation is called symmetric if $K(t, x) = K(x, t)$. An equation with a symmetric kernel is called symmetric.

The following are properties of a symmetric kernel:

1. A symmetric kernel has at least one characteristic number;
2. All characteristic numbers of a symmetric kernel are real;
3. The eigenfunctions corresponding to different characteristic numbers are orthogonal to each other.

In addition, as for any kernel, each characteristic number corresponds to a finite number of linearly independent eigenfunctions.

More details about symmetric integral equations can be found in [1–5]. In this chapter, the main methods for solving symmetric integral equations are presented. More examples and exercises can be found in [6].

Let the characteristic number λ_0 of the symmetric kernel $K(x, t)$ corresponds to m finite number of linearly independent eigenfunctions

$$\varphi_1(x), \varphi_2(x), \ldots, \varphi_m(x).$$

They can be orthogonalized, that is, m new functions

$$\psi_1(x), \psi_2(x), \ldots, \psi_m(x)$$

can be constructed, that are mutually orthogonal to each other:

$$(\psi_i, \psi_j) = 0 \text{ if } i \neq j, (\psi_i, \psi_i) = \|\psi_i\|^2 \neq 0.$$

As is known from linear algebra, the process of orthogonalization is as follows:

© The Author(s), under exclusive license to Springer Nature Singapore Pte Ltd. 2022
V. Ryzhov et al., *Modern Methods in Mathematical Physics*,
https://doi.org/10.1007/978-981-19-4915-9_4

$$\psi_1(x) \equiv \varphi_1(x);$$

$$\psi_2(x) = \varphi_2(x) - \frac{(\varphi_2, \psi_1)}{\|\psi_1\|^2}\psi_1(x);$$

$$\psi_3(x) = \varphi_3(x) - \frac{(\varphi_3, \psi_1)}{\|\psi_1\|^2}\psi_1(x) - \frac{(\varphi_3, \psi_2)}{\|\psi_2\|^2}\psi_2(x);$$

$$\cdots$$

$$\psi_m(x) = \varphi_m(x) - \sum_{i=1}^{m-1}\frac{(\varphi_m, \psi_i)}{\|\psi_i\|^2}\psi_1(x).$$

Since the functions $\psi_1(x)$, $\psi_2(x)$, ..., $\psi_m(x)$ are linear combinations of eigenfunctions $\varphi_1(x)$, $\varphi_2(x)$, ..., $\varphi_m(x)$, they are also eigenfunctions of the kernel corresponding to the characteristic number λ_0. Due to their mutual orthogonality, they are also linearly independent.

Example 4.1 Functions $1, x, x^2, x^3$ are linearly independent in $[-1, 1]$. Construct orthonormal functions from them.

Solution

$$\psi_1(x) = 1; \quad \|\psi_1\| = \sqrt{\int_{-1}^{1}\psi_1^2(x)dx} = \sqrt{\int_{-1}^{1}1^2dx} = \sqrt{2};$$

$$\psi_2(x) = x - \frac{(x, \psi_1)}{\|\psi_1\|^2}\psi_1(x) = x - \frac{1}{2}\int_{-1}^{1}xdx = x; \quad \|\psi_2\| = \sqrt{\int_{-1}^{1}x^2dx} = \sqrt{\frac{2}{3}};$$

$$\psi_3(x) = x^2 - \frac{(x^2, \psi_1)}{\|\psi_1\|^2}\psi_1(x) - \frac{(x^2, \psi_2)}{\|\psi_2\|^2}\psi_2(x) =$$

$$x^2 - \frac{1}{2}\int_{-1}^{1}x^2(x)dx - \frac{3}{2}\int_{-1}^{1}x^3(x)dx = x^2 - \frac{1}{3};$$

$$\|\psi_3\| = \sqrt{\int_{-1}^{1}\left(x^2 - \frac{1}{3}\right)^2dx} = \frac{2}{3}\sqrt{\frac{2}{5}};$$

$$\psi_4(x) = x^3 - \frac{(x^3, \psi_1)}{\|\psi_1\|^2}\psi_1(x) - \frac{(x^3, \Psi_2)}{\|\psi_2\|^2}\psi_2(x) - \frac{(x^3, \psi_3)}{\|\psi_3\|^2}\psi_3(x) =$$

$$x^3 - \frac{1}{2}\int_{-1}^{1}x^3dx - \frac{3}{2}x\int_{-1}^{1}x^4dx -$$

$$\frac{45}{8}\left(x^2 - \frac{1}{3}\right) \int\limits_{-1}^{1} x^3 \left(x^2 - \frac{1}{3}\right) dx = x^3 - \frac{3}{5}x$$

$$\|\psi_4\| = \sqrt{\int\limits_{-1}^{1} \left(x^3 - \frac{3}{5}x\right)^2 dx} = \frac{2}{5}\sqrt{\frac{2}{7}}.$$

The new functions $1, x, x^2 - \frac{1}{3}, x^3 - \frac{3}{5}x$ are mutually orthogonal in $[-1; 1]$. We obtain the orthonormalized functions dividing them by norms:

$$\frac{1}{\sqrt{2}}; \sqrt{\frac{3}{2}}x; \frac{1}{2}\sqrt{\frac{5}{2}}(3x^2 - 1); \frac{1}{2}\sqrt{\frac{7}{2}}(5x^3 - 3x).$$

This example can be solved in *Wolfram Mathematica* [7] using the built-in **Orthonormalize**. We will write a function for this.

```
In [1]:= Clear[Orthonormalize]
         Orthonormalize[functions_,{x_,a_,b_}]:=
           Orthogonalize[functions,Integrate[#1
#2,{x,a,b}]&]
```

We apply the written function to the given sequence.

```
In [3]:= funcSeq={1,x,x^2,x^3};
         a=-1;
         b=1;
         Orthonormalize[funcSeq,{x,a,b}]
```

$$\text{Out[6]} := \left\{\frac{1}{\sqrt{2}}, \sqrt{\frac{3}{2}}x, \frac{3}{2}\sqrt{\frac{5}{2}}\left(-\frac{1}{3} + x^2\right), \frac{5}{2}\sqrt{\frac{7}{2}}\left(-\frac{3x}{5} + x^3\right)\right\}$$

Answer

$$\frac{1}{\sqrt{2}}; \sqrt{\frac{3}{2}}x; \frac{1}{2}\sqrt{\frac{5}{2}}(3x^2 - 1); \frac{1}{2}\sqrt{\frac{7}{2}}(5x^3 - 3x).$$

So, linearly independent eigenfunctions of a symmetric kernel corresponding to one eigenvalue can be orthonormalized. By property 3, applying the orthogonalization process to each set of eigenfunctions, one can obtain an orthonormal system of kernel eigenfunctions corresponding to all characteristic numbers. Since there is a

countable set of all characteristic numbers, and a finite number of functions correspond to each characteristic number, we obtain a countable set of orthonormalized
functions.

4.2 Representation of the Solution as Expansion in Terms of Orthonormal Eigenfunctions of a Symmetric Kernel

Consider the symmetric Fredholm equation of the second kind

$$\varphi(x) = \lambda \int_a^b K(x, t)\varphi(t)\mathrm{d}t + f(x) \tag{4.1}$$

where $K(x, t) = K(t, x)$. Let us assume that we know the characteristic numbers and
eigenfunctions of the kernel. We arrange all the characteristic values of a symmetric
kernel in order (by property 2, they are real), repeating each number as many times
as its rank: $\lambda_1, \lambda_2, \ldots, \lambda_n, \ldots$ Then, each number corresponds to one independent
function, and all functions can be considered orthonormal $\varphi_1(x), \varphi_2(x), \varphi_n(x), \ldots$
The solution of the symmetric equation can be obtained in the form of a Fourier
series in the orthogonal system of eigenfunctions $\{\varphi_n(x)\}$.

There are three possible cases described in the next paragraphs.

(1) Parameter λ of Eq. (4.1) is not a characteristic number, i.e., $\lambda \neq \lambda_n, n = 1, 2, \ldots$ Then, the equation has a unique solution, which can be written as the
kernel eigenfunction Fourier expansion

$$\varphi(x) = f(x) + \lambda \sum_{n=1}^{\infty} \frac{(f, \varphi_n)}{(\lambda_n - \lambda)\|\varphi_n\|^2}\varphi_n(x). \tag{4.2}$$

(2) Parameter λ of Eq. (4.1) is a characteristic number. Then, it occurs in the
sequence $\lambda_1, \lambda_2, \ldots, \lambda_n, \ldots r$ times, if its rank is r. For example, $\lambda = \lambda_j = \lambda_{j+1} = \cdots = \lambda_{j+r-1}$. In this case, the equation is solvable only if the free term
is $f(x)$ orthogonal to the eigenfunctions corresponding to the characteristic
number, i.e., if

$$\left(f, \varphi_j\right) = \left(f, \varphi_{j+1}\right) = \cdots = \left(f, \varphi_{j+z}\right) = 0. \tag{4.3}$$

This conclusion coincides with the result of III Fredholm theorem, since for a
symmetric kernel the original and associated homogeneous equations are the same.

If condition (4.3) is satisfied, then the equation has an infinite set of solutions.
The general solution of the equation can be written as

$$\varphi(x) = f(x) + \lambda \sum_{n=1}^{\infty} \frac{(f, \varphi_n)}{(\lambda_n - \lambda)\|\varphi_n\|^2} \varphi_n(x)$$
$$+ c_1\varphi_j(x) + c_2\varphi_{j+1}(x) + \cdots + c_{r+1}\varphi_{j+r}(x), \tag{4.4}$$

where $c_1, c_2, \ldots, c_{r+1}$ are arbitrary constants.

(3) Parameter λ of Eq. (4.1) is a characteristic number, but the free term $f(x)$ is not orthogonal to the eigenfunctions corresponding to λ, so condition (4.3) is not satisfied. In this case, the integral Eq. (4.1) has no solutions.

So, the solution to the symmetric equation can be obtained in the form of (4.2) or (4.4), or does not exist at all. The problem of solving a symmetric equation is thus reduced to determining the characteristic numbers and eigenfunctions of the kernel.

The characteristic numbers and eigenfunctions can be found by considering a corresponding homogeneous integral equation

$$\varphi(x) = \lambda \int_a^b K(x, t)\varphi(t)dt. \tag{4.5}$$

Recall that the characteristic numbers are the values of the parameter λ, for which the homogeneous Eq. (4.5) has an infinite set of solutions and the eigenfunctions are only linearly independent solutions from this set.

If the symmetric kernel is degenerate, then its characteristic numbers can be found as the roots of the Fredholm determinant according to the algorithm described in Chap. 3. Since every degenerate kernel has a finite number of characteristic numbers, the series on the right-hand sides of formulas (4.2) and (4.4) are replaced by finite sums.

Sometimes, it is more convenient to search for characteristic numbers and eigenfunctions of a symmetric kernel using the method of differentiation (see 1.4), i.e., reducing the homogeneous equation to a boundary value problem for the corresponding differential equation.

In some cases, an inhomogeneous symmetric integral equation can be reduced to an inhomogeneous boundary value problem for the corresponding differential equation.

Example 4.2 Solve the following equation using the kernel eigenfunctions expansion.

$$\varphi(x) - \lambda \int_0^{\pi} \cos(x + t)\varphi(t)dt = \cos 2x.$$

Solution

The kernel of the equation is symmetric, since $\cos(x + t) = \cos(t + x)$. It is also degenerate (see Example 3.12). To determine its characteristic numbers and eigenfunctions, consider the homogeneous equation

$$\varphi(x) = \lambda \int\limits_0^\pi \cos(x + t)\varphi(t)dt.$$

From Example 3.12, we already know that its characteristic numbers are $\lambda_1 = 2/\pi, \lambda_2 = -2/\pi$ and they comply with their eigenfunctions $\varphi_1(x) = \cos x$, $\varphi_2(x) = \sin x$. These functions are orthogonal according to property 3 of the symmetric kernel.

The unique solution of the original equation for $\lambda \neq \pm 2/\pi$ can be expressed by the formula (4.2):

$$\varphi(x) = f(x) + \lambda \frac{(f, \varphi_1)}{\|\varphi_1\|^2(\lambda_1 - \lambda)}\varphi_1(x) + \lambda \frac{(f, \varphi_2)}{\|\varphi_2\|^2(\lambda_2 - \lambda)}\varphi_2(x).$$

Here,

$$\|\varphi_1\|^2 = \int\limits_0^\pi \cos^2 x dx = \frac{\pi}{2};$$

$$\|\varphi_2\|^2 = \int\limits_0^\pi \sin^2 x dx = \frac{\pi}{2};$$

$$(f, \varphi_1) = \int\limits_0^\pi \cos 2x \cdot \cos x dx = 0;$$

$$(f, \varphi_2) = \int\limits_0^t \cos 2x \cdot \sin x dx = -\frac{2}{3}.$$

Then, for $\lambda \neq \pm\frac{2}{\pi}$, equation has a unique solution:

$$\varphi(x) = \cos 2x + \frac{4\lambda}{3(2 + \pi\lambda)}\sin x.$$

If $\lambda = \lambda_1 = 2/\pi$, then the equation has an infinite number of solutions, as $f(x)$ is orthogonal to eigenfunction $\varphi_1(x)$:

$$(f, \varphi_1) = 0.$$

All solutions can be expressed by the formula:

$$\varphi(x) = c \cos x + \cos 2x + \frac{2}{3\pi} \sin x,$$

where c is an arbitrary constant.

If $\lambda = \lambda_2 = -2/\pi$, then the equation has no solutions, because $f(x)$ is not orthogonal to eigenfunction $\varphi(x)$:

$$(f, \varphi_2) = -2/3 \neq 0.$$

Let us develop a scheme for solving equations with degenerate symmetric kernels in *Wolfram Mathematica* [7].

Let us write a number of auxiliary functions based on examples from Chap. 3. We will slightly change the function to find the eigenfunctions of the equation (it will be auxiliary in this problem) and also define a function for constructing a solution.

```
In [1]:= Clear[EigenValuesFunctions]
        EigenValuesFunctions[ak_,bk_,{a_,b_},x_]:=Block[
         {n,cList,coefMatrix,eigenvalues,linearsystems,
           linearsolutions,eigenfunctionsconst,eigenfunc-
tions,t},
           n=Length@ak;
           cList=C/@Range[n];
           coefMatrix=Inte-
grate[Outer[Times,bk[t],ak[t]],{t,a,b}];
           eigenvalues=λ/.
             Solve[Det[(IdentityMatrix[n]-λ
coefMatrix)]==0,λ];
           linearsystems=Thread[(IdentityMatrix[n]-
                    # coefMatrix).cList==0]&/@eigenvalues;
           linearsolutions=ToRules@Re-
duce[#,cList]&/@linearsystems;
           eigenfunctionsconst=cList.ak[t]/.
                linearsolutions//FullSimplify;
           eigenfunctions=If[ MatchQ[#,Times[__]],Replace[#,
                    Long-
est[const__]func_:>func/;FreeQ[{const},t]],#]&/@
                eigenfunctionsconst;
           Transpose@{eigenvalues,eigenfunctions}/.t->x
    ]
In[3]:= Clear[SymmetricDegenerate]
        SymmetricDegener-
ate[ak_,bk_,{a_,b_},f_,λ_,x_]:=Block[
         {eigens,scalprods,n,λList,sols,conds,t},
           eigens=EigenValuesFunctions[ak,bk,{a,b},t];
           scalprods=Integrate[f[t]#,{t,a,b}]&/@ei-
gens[[All,2]];
           n=Length@eigens;
           λList=Flatten@{λ,eigens[[All,1]]};
           sols=f[t]+# Sum[If[!(#===eigens[[k,1]]),
```

```
                scalprods[[k]]/(eigens[[k,1]]-#)/
                   Integrate[eigens[[k,2]]^2,{t,a,b}]
                eigens[[k,2]],0]+
             If[#===eigens[[k,1]]&&scalprods[[k]]==0,
             C[k]eigens[[k,2]]/#,0]+
             If[#===eigens[[k,1]]&&scalprods[[k]]!=0,
             Indeterminate,0],{k,n}]&/@λList
        //FullSimplify//Expand;
           conds={And@@Thread[λ!=eigens[[All,1]]]}~Join~
             Thread[λ==eigens[[All,1]]];
           Piecewise[Ta-
ble[{sols[[k]],conds[[k]]},{k,n+1}]]/.t->x
             ]
```

Set the initial data and check how these functions work. At the same time, we get a solution of the equation.

```
In [5]:= Clear[ak,bk,a,b,f,φ,λ,x]
         ak=Function[x,{Cos[x],-Sin[x]}];
         bk=Function[t,{Cos[t],Sin[t]}];
         f=Function[x,Cos[2x]];
         a=0;
         b=π;
         eqn=φ[x]==f[x]+λ Integrate[ak[x].bk[t]
φ[t],{t,a,b}]
```

$$\text{Out[11]:= } \phi[x] == \text{Cos}[2x] + \lambda \int_0^\pi (\text{Cos}[t]\text{Cos}[x] - \text{Sin}[t]\text{Sin}[x])\phi[t]dt$$

```
In [12]:= EigenValuesFunctions[ak,bk,{a,b},t]
```

$$\text{Out[12]:= } \left\{ \left\{ -\tfrac{2}{\pi}, \text{Sin}[t] \right\}, \left\{ \tfrac{2}{\pi}, \text{Cos}[t] \right\} \right\}$$

The **EigenValuesFunctions** function returns pairs {*characteristic number, eigenfunction*}. As seen, we get the expected result. Now, we will construct the solution and check it by substituting it into the equation.

```
In [13]:= sol=SymmetricDegenerate[ak,bk,{a,b},f,λ,x]
```

$$\text{Out[13]:= } \begin{cases} \text{Cos}[2x] + \frac{4\lambda\text{Sin}[x]}{6+3\pi\lambda} & \lambda \neq -\tfrac{2}{\pi} \&\& \lambda \neq \tfrac{2}{\pi} \\ \text{Indeterminate} & \lambda == -\tfrac{2}{\pi} \\ c_2\text{Cos}[x] + \text{Cos}[2x] + \frac{2\text{Sin}[x]}{3\pi} & \lambda == \tfrac{2}{\pi} \\ 0 & \text{True} \end{cases}$$

```
In [14]:= eqn/.φ->Function[x,sol[[1,1,1]]]//Evaluate]//Full-
```

```
Simplify
            eqn/.φ->Function[x,sol[[1,3,1]]//Evaluate]/.λ-
>2/π//
            FullSimplify
Out[14]:= True
Out[15]:= True
```

We see that the results coincide with those obtained by manual calculations.

Answers

1. $\varphi(x) = \cos 2x + \frac{4\lambda}{3(2+\pi\lambda)} \sin x$, if $\lambda \neq \pm\frac{2}{\pi}$;
2. $\varphi(x) = c \cos x + \cos 2x + \frac{2}{3\pi} \sin x$, if $\lambda = \frac{2}{\pi}$, where c—is an arbitrary constant;
3. no solutions if $\lambda = -\frac{2}{\pi}$

Example 4.3 Solve the following equation using the kernel eigenfunctions expansion

$$\varphi(x) - \lambda \int_0^\pi \cos(x + t)\varphi(t)\mathrm{d}t = \sin 3x.$$

Solution
This is the equation from the previous example, but with a different free term. Let us set the initial data and use the function from the previous example.

```
In [16]:= Clear[ak,bk,a,b,f,φ,λ,x]
          ak=Function[x,{Cos[x],-Sin[x]}];
          bk=Function[t,{Cos[t],Sin[t]}];
          f=Function[x,Sin[3x]];
          a=0;
          b=π;
          eqn=φ[x]==f[x]+λ Integrate[ak[x].bk[t]
φ[t],{t,a,b}]
```

Out$[22] := \phi[x] == \lambda \int_0^\pi (\mathrm{Cos}[t]\mathrm{Cos}[x] - \mathrm{Sin}[t]\mathrm{Sin}[x])\phi[t]dt + \mathrm{Sin}[3x]$

```
In [23]:= sol=SymmetricDegenerate[ak,bk,{a,b},f,λ,x]
```

Out$[23] := \begin{cases} \mathrm{Sin}[3x] & \lambda \neq -\frac{2}{\pi}\&\&\lambda \neq \frac{2}{\pi} \\ c_1\mathrm{Sin}[x] + \mathrm{Sin}[3x] & \lambda == -\frac{2}{\pi} \\ c_2\mathrm{Cos}[x] + \mathrm{Sin}[3x] & \lambda == \frac{2}{\pi} \\ 0 & \text{True} \end{cases}$

```
In [24]:= eqn/.φ->Function[x,sol[[1,1,1]]//Evaluate]//Full-
```

```
Simplify
         eqn/.φ->Function[x,sol[[1,2,1]]//Evaluate]/.λ->-
2/π//
         FullSimplify
         eqn/.φ->Function[x,sol[[1,3,1]]//Evaluate]/.λ-
>2/π//
         FullSimplify
Out[24]:= True
Out[25]:= True
Out[26]:= True
```

Answers

1. $\varphi(x) = \sin 3x$, if $\lambda \neq \pm\frac{2}{\pi}$;
2. $\varphi(x) = c \cos x + \sin 3x$, if $\lambda = \frac{2}{\pi}$, where c—is an arbitrary constant;
3. $\varphi(x) = c \sin x + \sin 3x$, if $\lambda = -\frac{2}{\pi}$, where c—is an arbitrary constant;

Example 4.4 Solve the equation

$$\varphi(x) - \lambda \int_0^1 K(x,t)\varphi(t)\mathrm{d}t = \cos \pi x,$$

where

$$K(x,t) = \begin{cases} t(x+1), \ 0 \leqslant x \leqslant t \\ x(t+1), \ t \leqslant x \leqslant 1 \end{cases}.$$

Solution
The kernel of the equation is symmetric, since $K(x,t) = K(t,x)$. To find the characteristic numbers and eigenfunctions, consider the corresponding homogeneous equation:

$$\varphi(x) = \lambda \int_0^1 K(x,t)\varphi(t)\mathrm{d}t.$$

We divide the integration interval $[0, 1]$ into two sections and write on them the expressions for the kernel $K(x,t)$ and obtain

$$\varphi(x) = \lambda x \int_0^x (t+1)\varphi(t)\mathrm{d}t + \lambda(x+1) \int_x^1 t\varphi(t)\mathrm{d}t.$$

Twice differentiate this equality with respect to x, one obtains

$$\varphi'(x) = \lambda \int_0^x (t+1)\varphi(t)dt + \lambda x \cdot (x+1)\varphi(x)$$

$$+ \lambda \int_x^1 t\varphi(t)dt - \lambda(x+1)x\varphi(x).$$

Simplifying, we get

$$\varphi'(x) = \lambda \int_0^x (t+1)\varphi(t)dt + \lambda \int_x^1 t\varphi(t)dt.$$

and finally

$$\varphi''(x) = \lambda(x+1)\varphi(x) - \lambda x\varphi(x).$$

Get a differential equation of the second order

$$\varphi''(x) - \lambda\varphi(x) = 0.$$

To determine the boundary conditions, we find the values $\varphi(x)$ and $\varphi'(x)$ at the ends of the interval:

$$\varphi(0) = \lambda \int_0^1 t\phi(t)dt, \varphi(1) = \lambda \int_0^1 (t+1)\varphi(t)dt,$$

$$\varphi'(0) = \varphi(0), \varphi'(1) = \varphi(1).$$

So, a homogeneous integral equation is reduced to a boundary value problem for a second-order linear differential equation

$$\varphi''(x) - \lambda\varphi(x) = 0, \varphi'(0) = \varphi(0), \varphi'(1) = \varphi(1).$$

To solve the obtained problem, various values of λ must be considered:

(1) $\lambda = 0$. The equation takes the form

$$\varphi''(x) = 0.$$

Hence

$$\varphi'(x) = c_1,$$

$$\varphi(x) = c_1 x + c_2.$$

Using the boundary conditions

$$c_1 = c_2,$$
$$c_1 + c_2 = c_1,$$

so

$$c_1 = c_2 = 0.$$

Therefore, $\varphi(x) = 0$ is the unique solution to the equation. The number $\lambda = 0$ is not characteristic;

(2) $\lambda > 0$. We solve the linear equation

$$\varphi''(x) - \lambda\varphi(x) = 0$$

using the characteristic equation

$$k^2 - \lambda = 0,$$

whence $k_{1,2} = \pm\sqrt{\lambda}$ and general solution of the equation

$$\varphi(x) = c_1 e^{\sqrt{\lambda}x} + c_2 e^{-\sqrt{\lambda}x}.$$

Find

$$\varphi'(x) = c_1\sqrt{\lambda}e^{\sqrt{\lambda}x} - c_2\sqrt{\lambda}e^{-\sqrt{\lambda}x}.$$

Substituting the boundary conditions, we obtain

$$\begin{cases} c_1 + c_2 = (c_1 - c_2)\sqrt{\lambda}; \\ c_1 e^{\sqrt{\lambda}} + c_2 e^{-\sqrt{\lambda}} = \sqrt{\lambda}\left(c_1 e^{\sqrt{\lambda}} - c_2 e^{-\sqrt{\lambda}}\right) \end{cases}$$

or

$$\begin{cases} c_1\left(1 - \sqrt{\lambda}\right) + c_2\left(1 + \sqrt{\lambda}\right) = 0; \\ c_1 e^{\sqrt{\lambda}}\left(1 - \sqrt{\lambda}\right) + c_2 e^{-\sqrt{\lambda}}\left(1 + \sqrt{\lambda}\right) = 0. \end{cases}$$

The uniqueness of the solution of this system depends on the determinant

$$D(\lambda) = \begin{vmatrix} 1 - \sqrt{\lambda} & 1 + \sqrt{\lambda} \\ e^{\sqrt{\lambda}}\left(1 - \sqrt{\lambda}\right) & e^{-\sqrt{\lambda}}\left(1 + \sqrt{\lambda}\right) \end{vmatrix} = 2(\lambda - 1)\sinh\sqrt{\lambda}.$$

If $D(\lambda) \neq 0$, then the system has a unique solution $c_1 = c_2 = 0$; hence, $\varphi(x) = 0$ is a unique solution of the homogeneous equation. If $D(\lambda) = 0$, then the system (4.6) has an infinite set of solutions and the homogeneous integral equation has nontrivial solutions. Therefore, the roots of the equation $D(\lambda) = 0$ are characteristic numbers. Then, if $\lambda > 0$, there is a unique characteristic number $\lambda_0 = 1$.

From the system of algebraic equations for $\lambda = 1$, we get $c_2 = 0$, c_1 is arbitrarily. Consequently, all solutions of the homogeneous equation for $\lambda = 1$ have the form

$$\varphi(x) = c_1 e^x,$$

where c_1 is an arbitrary constant. The eigenfunction corresponding to the number $\lambda_0 = 1$ can be taken as

$$\varphi_0(x) = e^x.$$

(3) $\lambda < 0$. The characteristic equation

$$k^2 - \lambda = 0$$

gives

$$k_{1,2} = \pm i\sqrt{|\lambda|}.$$

General solution of a differential equation

$$\varphi(x) = c_1 \cos\sqrt{|\lambda|}x + c_2 \sin\sqrt{|\lambda|}x.$$

Differentiate it

$$\varphi'(x) = -c_1\sqrt{|\lambda|}\sin\sqrt{|\lambda|}x + c_2\sqrt{|\lambda|}\cos\sqrt{|\lambda|}x.$$

Substituting the boundary conditions, we obtain a system for determining c_1 and c_2:

$$\begin{cases} c_1 - \sqrt{|\lambda|}c_2 = 0; \\ c_1\left(\cos\sqrt{|\lambda|} + \sqrt{|\lambda|}\sin\sqrt{|\lambda|}\right) + c_2\left(\sin\sqrt{|\lambda|} - \sqrt{|\lambda|}\cos\sqrt{|\lambda|}\right) = 0. \end{cases}$$

Determinant of the system

$$D(\lambda) = \sin\sqrt{|\lambda|}(1 + |\lambda|).$$

The roots of the equation $D(\lambda) = 0$ are characteristic numbers:

$$\sin \sqrt{|\lambda|} = 0 \Leftrightarrow \sqrt{|\lambda|} = \pi n,$$
$$n = 1, 2, \ldots \Leftrightarrow |\lambda| = \pi^2 n^2 \Leftrightarrow \lambda = -\pi^2 n^2 (\lambda < 0).$$

So, if $\lambda < 0$, there is an infinite set of characteristic numbers $\lambda_n = -n^2 \pi^2$, $n = 1, 2, \ldots$. System (4.7) for $\lambda = \lambda_n$ gives $c_1 = n\pi c_2$. Then, the solution to the boundary value problem takes the form

$$\varphi(x) = c_2(\pi n \cos \pi n x + \sin \pi n x),$$

where C_2 is an arbitrary constant.

The eigenfunctions corresponding to numbers $\lambda_n (n = 1, 2, \ldots)$ can be taken in the form

$$\varphi_n(x) = \pi n \cos \pi n x + \sin \pi n x.$$

Each eigenvalue λ_n corresponds to a unique independent eigenfunction $\varphi_n(x)$ (the rank of the number λ_n is equal to one). Consequently, all $\varphi_n(x), n = 0, 1, 2, \ldots$, including $\varphi_0(x) = e^x$, are orthogonal.

For $\lambda \neq 1$, $\lambda \neq -n^2 \pi^2 (n = 1, 2, \ldots)$, there is a unique solution of the integral equation, which can be written [see (4.2)] in the form

$$\varphi(x) = \cos \pi x + \frac{(f, \varphi_0)\lambda}{\|\varphi_0\|^2 (1 - \lambda)} \varphi_0(x) + \lambda \sum_{n=1}^{\infty} \frac{(f, \varphi_n)}{\|\varphi_n\|^2 (\lambda_n - \lambda)} \varphi_n(x).$$

Now, we find the norms and the scalar products

$$\|\varphi_0\|^2 = \int_0^1 e^{2x} \mathrm{d}x = \frac{1}{2} e^{2x} \Big|_0^1 = \frac{1}{2}(e^2 - 1);$$

$$\|\varphi_n\|^2 = \int_0^1 (\sin n\pi x + \pi n \cos n\pi x)^2 \mathrm{d}x = \frac{1 + \pi^2 n^2}{2};$$

$$(f, \varphi_0) = \int_0^1 e^x \cos \pi x \mathrm{d}x = -\frac{1 + e}{\pi^2 + 1};$$

$$(f, \varphi_n) = \int_0^1 \cos \pi x (\sin n\pi x + n\pi \cos n\pi x) \mathrm{d}x$$

$$= \frac{1}{2} \int_0^1 [\sin(n+1)\pi x + \sin(n-1)\pi x] dx$$

$$+ \frac{\pi n}{2} \int_0^1 [\cos(n+1)\pi x + \cos(n-1)\pi x] dx.$$

Here, it is necessary to consider the cases $n = 1$ and $n \neq 1$ separately.

If $n = 1$:

$$(f, \varphi_1) = \frac{1}{2} \int_0^1 \sin 2\pi x dx + \frac{\pi}{2} \int_0^1 (\cos 2\pi x + 1) dx = \frac{\pi}{2}.$$

If $n \neq 1$

$$(f, \varphi_n) = \frac{1}{2(n+1)\pi} [-\cos(n+1)\pi x] \Big|_0^1 + \frac{1}{2\pi(n-1)} [-\cos\pi(n-1)x] \Big|_0^1$$

$$+ \frac{\pi n}{2(n+1)\pi} [\sin(n+1)\pi x] \Big|_0^1 + \frac{\pi n}{(n-1)\pi} [\sin(n-1)\pi x] \Big|_0^1$$

$$= \begin{cases} 0, & n - \text{odd}; \\ \frac{2n}{\pi(n^2-1)}, & n - \text{even}. \end{cases}$$

Denoting even numbers $n = 2k$, we get

$$(f, \varphi_{2k}) = \frac{4k}{\pi(4k^2 - 1)}, \quad (f, \varphi_{2k+1}) = 0, \quad k = 1, 2, \ldots$$

The unique solution to the integral equation for a regular $\lambda (\lambda \neq 1, \ \lambda \neq -n^2 x^2, \ n = 1, 2, \ldots)$ takes the form

$$\varphi(x) = \cos \pi x - \frac{2\lambda}{(1-\lambda)(\pi^2+1)(e-1)} e^x$$

$$- \frac{\pi \lambda}{(1+\pi^2)(\pi^2+2)} (\sin \pi x + \pi \cos \pi x)$$

$$- \frac{8\lambda}{\pi} \sum_{K=1}^{\infty} \frac{k}{(4k^2 - 1)(1 + 4\pi^2 k^2)(\lambda + 4\pi^2 k^2)} (\sin 2k\pi x + 2k\pi \cos 2k\pi x).$$

- If $\lambda = \lambda_0 = 1$ equation has no solution. since $(f, \varphi_0) \neq 0$,
- If $\lambda = \lambda_1 = -\pi^2$ equation also has no solution, since $(f, \varphi_1) \neq 0$,
- If $\lambda = \lambda_n$ for even n equation has no solution,

For the odd, n > 1 $(f, \varphi_n) = 0$. Therefore, if $= \lambda_{2k+1} = -(2k+1)^2\pi^2$, $k = 1, 2, \ldots$, the equation has an infinite number of solutions.

In accordance with (4.4), all solutions of the integral equation for $\lambda = -(2k+1)^2\pi^2$, $k > 0$ can be expressed in the form

$$\varphi(x) = \cos\pi x + \frac{2(2k+1)^2\pi^2}{\left(1+(2k+1)^2\pi^2\right)\left(\pi^2+1\right)(e-1)}e^x +$$

$$\frac{\pi(2k+1)^2}{\left(1+\pi^2\right)\left(1-(2k+1)^2\right)}(\sin\pi x + \pi\cos\pi x) +$$

$$\frac{8(2k+1)^2}{\pi}\sum_{m=1}^{\infty}\frac{m}{\left(4m^2-1\right)\left(1+4\pi^2m^2\right)\left(4m^2-(2k+1)^2\right)}(\sin 2m\pi x +$$

$$2m\pi\cos 2m\pi x) + c(\sin(2k+1)\pi x + (2k+1)\pi\cos(2k+1)\pi x),$$

where c is an arbitrary constant.

In some cases, an inhomogeneous symmetric integral equation can be reduced to an inhomogeneous boundary value problem for the corresponding differential equation.

Now let us try to solve this equation in *Wolfram Mathematica*. We can use the results from the examples that we solved using the differentiation method. Firstly, define the auxiliary functions.

```
In [1]:= Clear[PullOut]
         PullOut[s_]:=s//.
           {Integrate[f_+g_,it:{x_Symbol,__}]
             :>Integrate[f,it]+Integrate[g,it],
            Integrate[c_ f_.,it:{x_Symbol,__}]
             :>c Integrate[f,it]/;FreeQ[c,x]}
In[3]:= Clear[ExtractIntegrals]
         ExtractIntegrals[exp_]:=
           Union@Extract[exp,Position[exp,Integrate[_,_]]]
```

Set the initial data.

```
In [5]:= Clear[kernel,a,b,f,φ,λ,x,t]
         kernel=Piece-
wise[{{t(x+1),0<=x<=t},{(x(t+1),t<=x<=1}}];
         f=Cos[π x];
         a=0;
         b=1;
         {kernel2,kernel1}=kernel[[1,All,1]];
         eqn=φ[x]==f+λ Integrate[kernel1 φ[t],{t,a,x}]+
            λ Integrate[kernel2 φ[t],{t,x,b}]//PullOut
```

$$\text{Out}[11]:=\phi[x]==\text{Cos}[\pi x]+\text{kernel1}\lambda\int_0^x\phi[t]dt+\text{kernel2}\lambda\int_x^1\phi[t]dt$$

Now let us write a homogeneous equation and differentiate it two times.

```
In [12]:= homeqn=φ[x]==
              λ Integrate[kernel1 φ[t],{t,a,x}]+
              λ Integrate[kernel2 φ[t],{t,x,b}]//PullOut
          dhomeqn=D[homeqn,x]
          difeqn=D[dhomeqn,x]//Expand
```

$$\text{Out}[12]:=\phi[x]==(1+x)\lambda\int_x^1 t\phi[t]dt+x\lambda\int_0^x(1+t)\phi[t]dt$$

$$\text{Out}[13]:=\phi'[x]==\lambda\int_x^1 t\phi[t]dt+\lambda\int_0^x(1+t)\phi[t]dt$$

$$\text{Out}[14]:=\phi''[x]==\lambda\phi[x]$$

Further, we obtain the boundary conditions. To do this, we substitute in the homogeneous equation and its derivative the values $x=a$ and $x=b$ and then exclude the integrals.

```
In [15]:=valuesOnEnds=Flatten[{homeqn,dhomeqn}/.{{x->a},{x->b}}]
            initcond=List@@Eliminate[valuesOnEnds,
              ExtractIntegrals[valuesOnEnds]]
```

$$\text{Out}[15]:=\{\phi[0]==\lambda\int_0^1 t\phi[t]dt,\phi'[0]==\lambda\int_0^1 t\phi[t]dt,$$

$$\phi[1]==\lambda\int_0^1(1+t)\phi[t]dt,\phi'[1]==\lambda\int_0^1(1+t)\phi[t]dt\}$$

$$\text{Out}[16]:=\{\phi[0]==\phi'[0],\phi[1]==\phi'[1]\}$$

Now, solve the differential equation.

```
In [17]:= DSolveValue[{difeqn,initcond},φ[x],x]
```

$$
\text{Out[17]}:= \begin{cases} \dfrac{2c_1\left(\sqrt{\lambda}\cosh\left[x\sqrt{\lambda}\right]+\sinh[x\sqrt{\lambda}]\right)}{-1+\sqrt{\lambda}}\,(-1+\lambda)\sinh\left[\sqrt{\lambda}\right]==0 \\ 0 \qquad\qquad\qquad\qquad\qquad \textbf{True} \end{cases}
$$

We get the solution under the condition in the form of an equation. Here, it is necessary to take into account that *Mathematica* considers that values λ can be complex so we can refine the values λ when solving the equation using the **Assumptions** option. In this case, to obtain solutions for $\lambda < 0$ more efficient way would be to replace λ in the equation with $-\lambda$.

```
In [18]:= sol1=DSolveValue[{difeqn,initcond},φ[x],x,
              Assumptions->λ==0]
          sol2=DSolveValue[{difeqn/.λ->-λ,initcond},φ[x],x,
              Assumptions->λ>0]/.λ->-λ
          sol3=DSolveValue[{difeqn,initcond},φ[x],x,
              Assumptions->λ>0]
Out[18]:= 0
```

$$
\text{Out[19]}:=
$$

$$
\begin{cases} \left(\sqrt{\lambda}\left(\cos\left[x\sqrt{\lambda}\right]+\sin[x\sqrt{\lambda}]\right)\right) \quad \begin{array}{l} n\in\mathbb{Z}\&\&\left(\left(n\geq 1\&\&-\lambda==4n^2\pi^2\right)\|\right. \\ \left.\left(n\geq 0\&\&-\lambda==+\pi^2+4n\pi^2+4n^2\pi^2\right)\right) \end{array} \\ 0 \qquad\qquad\qquad\qquad\qquad\qquad\qquad \textbf{True} \end{cases}
$$

$$
\text{Out[20]}:= \begin{cases} \dfrac{2c_1\left(\sqrt{\lambda}\cosh\left[x\sqrt{\lambda}\right]+\sinh[x\sqrt{\lambda}]\right)}{-1+\sqrt{\lambda}} \quad \lambda==1 \\ 0 \qquad\qquad\qquad\qquad\qquad\quad \textbf{True} \end{cases}
$$

For $\lambda = 0$, we have a trivial solution. From the other two solutions, we obtain equations containing characteristic numbers and extract them. It can be seen that for $\lambda < 0$, there will be a countable set, since λ depends on the depends on the integer n. There are two cases: In the first, $n = 1, 2, 3, \ldots$, and in the second, $n = 0, 1, 2, \ldots$. Therefore, for the convenience, we separate the case for $n = 0$.

```
In   [21]:=  λeqns=Flatten[Extract[#/.n->n,Position[#,_.
λ==_]]&/@

                {sol2[[1,1,2]],sol3[[1,1,2]]}]
             eigenValues=λ/.Solve[#,λ][[1]]&/@λeqns;
             AppendTo[eigenValues,eigenValues[[2]]/.n->0]
//Column
```

$$\text{Out[21]} := \left\{ -\lambda == 4n^2\pi^2, -\lambda == \pi^2 + 4n\pi^2 + 4n^2\pi^2, \lambda == 1 \right\}$$

$$\text{Out[23]} := \begin{array}{c} -4n^2\pi^2 \\ -\pi^2 - 4n\pi^2 - 4n^2\pi^2 \\ 1 \\ -\pi^2 \end{array}$$

Now, you can substitute them back into the differential equation. In this case, the solutions will correspond to their eigenfunctions. We also define an auxiliary function that removes the constant before the function.

```
In [24]:= Clear[DropConstMultiplier]
          DropConstMultiplier[func_,x_]:=
          If[ MatchQ[func,Times[__]],
          Replace[func,Long-
est[c__]f_:>f/;FreeQ[{c},x]],func]
In[26]:= eigenFunctionsWithConst=
          FullSimplify[DSolveValue[{difeqn/.λ->#,in-
itcond},φ[x],x]&
                /@eigenValues,n ∈ Integers&&n >  = 0];
          eigenFunctions=DropConstMultiplier[#,x]&/@
          eigenFunctionsWithConst;
          eigenFunctions//Column
```

$$\text{Out[28]} := \begin{array}{c} 2n\pi\,\text{Cos}[2n\pi x] + \text{Sin}[2n\pi x] \\ \dfrac{(1+2n)\pi\,\text{Cos}[(1+2n)\pi x] + \text{Sin}[(1+2n)\pi x]}{e^x} \\ \pi\,\text{Cos}[\pi x] + \text{Sin}[\pi x] \end{array}$$

To obtain a solution, it remains to calculate the scalar products of the free term by the eigenfunctions. Since we have eigenfunctions which depend on natural n, first we calculate the integrals without this assumption to check that no special values are lost.

```
In [29]:= scalprods=Integrate[f #,{x,a,b}]&/@eigenFunctions
```

```
Out[29]:=
```

$$\left\{ \frac{4n\left(\cos[n\pi]^2 - n\pi\sin[2n\pi]\right)}{(-1+4n^2)\pi}, \frac{(1+2n)\sin[n\pi]((1+2n)\pi\cos[n\pi] + \sin[n\pi])}{2n(1+n)\pi}, -\frac{1+e}{1+\pi^2}, \frac{\pi}{2} \right\}$$

It can be seen that if $n = 1, 2, 3, \ldots$ no singularities arise, so we will help *Mathematica to* simplify the expressions by saying that n is integer. If you carry out such a check, then at the stage of simplifying the expression, *Mathematica* may throw out some special case.

```
In [30]:= scalprods=FullSimplify[scalprods,n∈Integers]
```

$$\text{Out}[30]:= \left\{ -\frac{4n}{\pi - 4n^2\pi}, 0, -\frac{1+e}{1+\pi^2}, \frac{\pi}{2} \right\}$$

As a result, we see that the results of calculating the characteristic numbers and eigenfunctions in *Wolfram Mathematica* completely coincided with the manual ones, so we will not repeat the same calculations. Let us check only that the obtained eigenfunctions are solutions of the homogeneous equation.

```
In [31]:= FullSimplify[homeqn/.
               φ->Function[x,Evaluate@eigenFunctions[[#]]]/.
               λ->eigenValues[[#]],n∈Integers]&/@Range@
            Length@eigenValues
Out[31]:= {True, True, True, True}
```

Let us build the graphs of the regular solution for different values of the parameter λ (Fig. 4.1). We need the norms of eigenfunctions to construct a solution.

```
In [32]:= eigenFunctionNorms=FullSimplify[Integr-
ate[#^2,{x,a,b}],
```

$$\text{n} \in \text{Integers}]\&/@\text{eigenFunctions}.$$
$$\text{Out}[32]:= \left\{ \tfrac{1}{2} + 2n^2\pi^2, \tfrac{1}{2}(1 + (\pi + 2n\pi)^2), \tfrac{1}{2}(-1 + e^2), \tfrac{1}{2}(1 + \pi^2) \right\}$$

```
In [33]:= Clear[regsol]
            regsol[λ_,m_]:=Evaluate@f+
              λ (scalprods[[3]]/eigenFunctionNorms[[3]]/
                   (eigenValues[[3]]-λ)Evaluate@
                   eigenFunctions[[3]]+
                scalprods[[4]]/eigenFunctionNorms[[4]]/
                   (eigenValues[[4]]-λ)Evaluate@
                   eigenFunctions[[4]]+
                Total@Table[scalprods[[1]]/
```

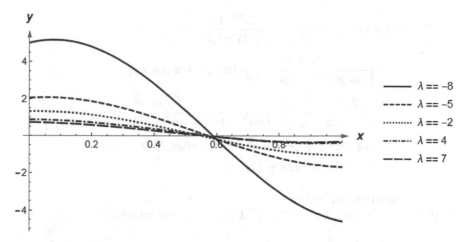

Fig. 4.1 Family of solutions of an integral equation for various regular values of the parameter λ (Example 4.4)

```
                    eigenFunctionNorms[[1]]/(eigenVal
ues[[1]]-λ)
                Evaluate@eigenFunctions[[1]],{n,m}])
  In[35]:= optplot={AxesStyle->Arrowheads[{0.0,0.025}],
          AxesLabel->{Style[x,Bold,Me-
dium],Style[y,Bold,Medium]}};
        Plot[regsol[λ,10]/.x->t/.λ->#//Evaluate,{t,a,b},
          Evaluate@optplot,PlotLegends-
>ToString/@Thread[λ==#]]&@
        Complement[Range[-8,10,3],{1}]
```

Answers

1. if $\lambda \neq 1$, $\lambda \neq -n^2 x^2$, $n = 1, 2, \ldots$

$$\varphi(x) = \cos \pi x - \frac{2\lambda}{(1-\lambda)(\pi^2+1)(e-1)} e^x$$

$$- \frac{\pi\lambda}{(1+\pi^2)(\pi^2+2)}(\sin \pi x + \pi \cos \pi x)$$

$$- \frac{8\lambda}{\pi} \sum_{K=1}^{\infty} \frac{k}{(4k^2-1)(1+4\pi^2 k^2)(\lambda + 4\pi^2 k^2)}$$

$$(\sin 2k\pi x + 2k\pi \cos 2k\pi x);$$

2. if $\lambda = -(2k+1)^2\pi^2, k = 1, 2, \ldots$

$$\varphi(x) = \cos \pi x + \frac{2(2k+1)^2\pi^2}{\left(1+(2k+1)^2\pi^2\right)\left(\pi^2+1\right)(e-1)}e^x +$$

$$\frac{\pi(2k+1)^2}{\left(1+\pi^2\right)\left(1-(2k+1)^2\right)}(\sin \pi x + \pi \cos \pi x);$$

$$+\frac{8(2k+1)^2}{\pi}\sum_{m=1}^{\infty}\frac{m}{\left(4m^2-1\right)\left(1+4\pi^2m^2\right)\left(4m^2-(2k+1)^2\right)}$$

$$(\sin 2m\pi x + 2m\pi \cos 2m\pi x) + c(\sin(2k+1)\pi x +$$

$$(2k+1)\pi \cos(2k+1)\pi x),$$

c is an arbitrary constant;

3. if $\lambda = 1, \lambda = -\pi^2, \lambda = -(2k)^2\pi^2, k = 1, 2, \ldots$ has no solution.

Example 4.5 Solve the integral equation by reducing to a differential equation

$$\varphi(x) - \lambda \int_0^1 K(x,t)\varphi(t)\mathrm{d}t = e^x$$

Where

$$K(x,t) = \frac{shx\,sh(t-1)}{sh1}, \quad 0 \le x \le t \qquad \frac{shx\,sh(x-1)}{sh1} \quad t \le x \le 1$$

Solution

The kernel of the equation is symmetric. We will find solution in *Wolfram Mathematica* according to the scheme analyzed in the previous example. However, let us try to reduce this equation to an inhomogeneous boundary value problem.

Firstly, set the initial data.

```
In [37]:= Clear[kernel,a,b,f,φ,λ,x,t]
          kernel=Piecewise[{{Sinh[x]Sinh[t-
1]/Sinh[1],0<=x<=t},
                {Sinh[t]Sinh[x-1]/Sinh[1],t<=x<=1}}];
          f=Exp[x];
          a=0;
          b=1;
          {kernel2,kernel1}=kernel[[1,All,1]];
          eqn=φ[x]==f+λ Integrate[kernel1 φ[t],{t,a,x}]+
                λ Integrate[kernel2 φ[t],{t,x,b}]//PullOut
```

$$\text{Out}[43]:= \quad \phi[x] == e^x - \lambda\,\text{Csch}[1]\left(\int_0^x \text{Sinh}[t]\phi[t]dt\right)\text{Sinh}[1-x]-$$

$$\lambda\,\text{Csch}[1]\left(\int_x^1 \text{Sinh}[1-t]\phi[t]dt\right)\text{Sinh}[x]$$

Now, we will not compose a homogeneous equation, but we will immediately differentiate the original one twice. In this case, in the second derivative, we will try to replace the sum of integrals by $\varphi(x)$.

```
In [44]:= deqn=D[eqn,x]
          difeqn=D[deqn,x]/.eqn[[2]]->φ[x]//FullSimplify
```

$$\text{Out}[44]:= \quad \phi'[x] == e^x - \lambda\,\text{Cosh}[x]\text{Csch}[1]\int_x^1 \text{Sinh}[1-t]\phi[t]dt$$

$$\lambda\,\text{Cosh}[1-x]\text{Csch}[1]\int_0^x \text{Sinh}[t]\phi[t]dt$$

$$\text{Out}[45]:= \quad (1+\lambda)\phi[x] == -\phi''[x]$$

Further, we obtain the boundary conditions.

```
In [46]:= valuesOnEnds=Flatten[{homeqn,dhomeqn}/.{{x->a},{x-
>a},{x->b}}]
          initcond=ToRules@Eliminate[valuesOnEnds,
          ExtractIntegrals[valuesOnEnds]]
```

$$\text{Out}[46]:= \quad \{\phi[0] == 1, \phi'[0] == 1 - \lambda\,\text{Csch}[1]\int_0^1 \text{Sinh}[1-t]\phi[t]dt,$$

$$\phi[1] == e, \phi'[1] == e + \lambda\,\text{Csch}[1]\int_0^1 \text{Sinh}[t]\phi[t]dt\}$$

$$\text{Out}[47]:= \quad \{\phi[0] == 1, \phi[1] == e\}$$

Note that now there is a factor $(\lambda + 1)$ in the differential equation at $\varphi(x)$. This means that the value $\lambda = -1$ will be transitional, we need to take this into account when solving the equation. We find solutions in the cases $\lambda = -1, \lambda > -1, \lambda < -1$, fixing the sign of multiplier $(\lambda + 1)$ using the **Abs** function. We can immediately solve the differential equation with the boundary conditions.

```
In [48] := λcrit=-1;
          {DSolveValue[{difeqn,initcond}/.λ->λcrit,φ[x],x],
           DSolveValue[{difeqn,initcond}/.
            λ-λcrit->Abs[λ-λcrit],v[x],x,
            Assumptions->λ>λcrit],
           DSolveValue[{difeqn,initcond}/.
            λ-λcrit->-Abs[λ-λcrit],φ[x],x,
            Assumptions->λ>λcrit]}//FullSimplify//Column
```

$$Out[49] := \begin{array}{c} 1+(-1+e)x \\ Csch[\sqrt{1+\lambda}](-Sinh[(-1+x)\sqrt{1+\lambda}]+eSinh[x\sqrt{1+\lambda}]) \\ Csc[\sqrt{1+\lambda}](-Sin[(-1+x)\sqrt{1+\lambda}]+eSin[x\sqrt{1+\lambda}]) \end{array}$$

But in this case, the risk of losing some special conditions under which the solution does not exist is extremely high. Therefore, first we will find the general form of the solution in each case.

```
In [50] := difsols=Flatten@{DSolve[difeqn/.λ->λcrit,φ[x],x],
             DSolve[difeqn/.λ-λcrit->Abs[λ-λcrit],φ[x],x,
              Assumptions->λ>λcrit],
             DSolve[difeqn/.λ-λcrit->-Abs[λ-λcrit],φ[x],x,
              Assumptions->λ>λcrit]}
          difsols //Column
```

$$Out[51] := \begin{array}{c} \phi \rightarrow \mathbf{Function}[\{x\}, c_1 + xc_2] \\ \phi \rightarrow \mathbf{Function}\left[\{x\}, e^{x\sqrt{1+\lambda}}c_1 + e^{-x\sqrt{1+\lambda}}c_2\right] \\ \phi \rightarrow \mathbf{Function}\left[\{x\}, c_1\mathbf{Cos}\left[x\sqrt{1+\lambda}\right] + c_2\mathbf{Sin}\left[x\sqrt{1+\lambda}\right]\right] \end{array}$$

And now, we will use the boundary conditions and compose systems of equations for the unknown coefficients. Let us find and investigate the determinants of these systems in order to obtain the constraints.

```
In [52] := coefsystems=Table[Coefficient[initcond[[i,1]]/.
                    #,C[j]],{i,2},{j,2}]&/@difsols;
          detsystems=Det/@coefsystems//FullSimplify
```

$$Out[53] := \left\{1, -2Sinh\left[\sqrt{1+\lambda}\right], Sin\left[\sqrt{-1-\lambda}\right]\right\}$$

```
In [54] := Reduce[detsystems[[1]]==0,λ]
          Reduce[{detsystems[[2]]==0,λ>λcrit},λ]
          Reduce[{detsystems[[3]]==0,λ<λcrit},λ]
Out[54] := False
Out[55] := False
```

$$\text{Out}[56] := c_1 \in \mathbb{Z} \&\& ((c_1 \geq 1 \&\& \lambda == -1 - 4\pi^2 c_1^2) ||$$
$$(c_1 \geq 0 \&\& \lambda == -1 - \pi^2 - 4\pi^2 c_1 - 4\pi^2 c_1^2))$$

False in the first two cases means that the determinant does not turn to zero. And, if $\lambda < -1$, we analyze the answer and get that the determinant turns to zero for $\lambda = -1 - \pi^2 n^2$. With this in mind, we express all the constants and substitute them in the corresponding solutions.

```
In [57]:= csols={Reduce[initcond/.difsols[[1]],{C[1],C[2]}],
                 Reduce[Flatten@{in-
itcond,λ>λcrit}/.difsols[[2]],
                        {C[1],C[2]}],
                 Reduce[Flatten@{in-
itcond,λ<λcrit}/.difsols[[3]],
                        {C[1],C[2]}]};
                 crules=Rule@@@Extract[#,Position[#,_.
C[_]==_]]&/@csols
```

$$\text{Out}[58] := \left\{\{c_1 \to 1, c_2 \to -1+e\}, \left\{c_1 \to \frac{-1+e^{1+\sqrt{1+\lambda}}}{-1+e^{2\sqrt{1+\lambda}}}, c_2 \to 1 - c_1\right\}\right.$$
$$\left. \left\{c_1 \to 1, c_2 \to \frac{1}{2}\text{Cot}\left[\frac{\sqrt{-1-\lambda}}{2}\right]\left(-1+e+\text{Tan}\left[\frac{\sqrt{-1-\lambda}}{2}\right]^2\right.\right.\right.$$
$$\left.\left.\left. +e\text{Tan}\left[\frac{\sqrt{-1-\lambda}}{2}\right]^2\right)\right\}\right\}$$

```
In [59]:= eqnsols=φ[x]/.difsols[[#]]//.crules[[#]]&/@
                 Range@Length@difsols//FullSimplify;
          eqnsols//Column
```

$$\text{Out}[60] := \begin{array}{c} 1 + (-1+e)x \\ \text{Csch}[\sqrt{1+\lambda}](-\text{Sinh}[(-1+x)\sqrt{1+\lambda}]+e\text{Sinh}[x\sqrt{1+\lambda}]) \\ \text{Csc}[\sqrt{-1-\lambda}](-\text{Sin}[(-1+x)\sqrt{-1-\lambda}]+e\text{Sin}[x\sqrt{-1-\lambda}]) \end{array}$$

Thus, we found all possible solutions. Let us check them by substituting into the original equation.

```
In [61]:= FullSimplify[eqn/.φ->Function[x,
                 Evaluate@eqnsols[[1]]]/.λ->-1]
          FullSimplify[eqn/.φ->Function[x,
                 Evaluate@eqnsols[[2]]],λ>-1]
          FullSimplify[eqn/.φ->Function[x,
```

```
                    Evaluate@eqnsols[[3]]],λ<-1]
Out[61]:= True
Out[62]:= True
Out[63]:= True
```

Finally, Figs. 4.2 Fig. 4.3 and 4.4 show the graphs of solutions for different values
λ .

```
In [64]:= Plot[eqnsols[[1]],{x,a,b},Evaluate@optplot]
          Plot[eqnsols[[2]]/.λ->#//Evaluate,{x,a,b},Evalu
ate@optplot,
          PlotLegends-
>ToString/@Thread[λ==#]]&@Range[0,5]
```

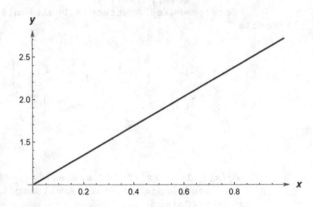

Fig. 4.2 Solution of the integral equation for the value of the parameter $\lambda = -1$ (Example 4.5)

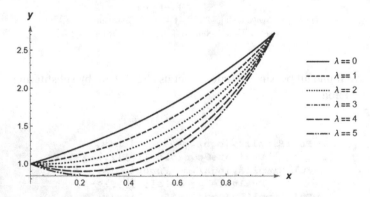

Fig. 4.3 Family of solutions of the integral equation for different values of the parameter λ in the case $\lambda > -1$ (Example 4.5)

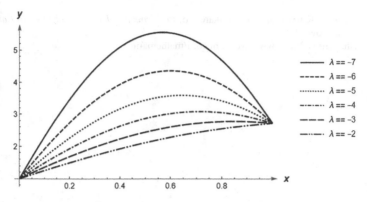

Fig. 4.4 Family of solutions of the integral equation for different values of the parameter λ in the case $\lambda < -1, \lambda \neq -1 - \pi^2 n^2$ (Example 4.5)

```
       Plot[eqnsols[[2]]/.λ->#//Evaluate,{x,a,b},Evalu-
ate@optplot,
             PlotLegends->ToString/@Thread[λ==#]]&@Range[-
7,-2]
```

Answers

1. if $\lambda = -1$ $\varphi(x) = (e - 1)x + 1$;
2. if $\lambda > -1$ $\varphi(x) = \frac{e \sinh \sqrt{\lambda+1}x + \sinh \sqrt{\lambda+1}(1-x)}{\sinh \sqrt{\lambda+1}}$;
3. if $\lambda < -1, \lambda \neq -1 - \pi^2 n^2$

$$\varphi(x) = \frac{e \sin \sqrt{-\lambda - 1}x + \sin \sqrt{-\lambda - 1}(1 - x)}{\sin \sqrt{-\lambda - 1}};$$

4. if $\lambda = -1 - \pi^2 n^2, n = 1, 2, \ldots$ there is no solution.

References

W. V. Lovitt, *Linear Integral Equations*, (Dover Publ., New York, 1950)

S. G. Mikhlin, *Linear Integral Equations*, (Courier Dover Publications, 2020)

S. G. Mikhlin, *Integral Equations and their Applications to Certain Problems in Mechanics, Mathematical Physics and Technology*, (Pergamon Press, 1964)

S. G. Mikhlin, *Integral Equations and their Applications to Certain Problems in Mechanics, Mathematical Physics and Technology*, (Pergamon Press, 1964)

P. P. Zabreyko, A. I. Koshelev, et al., Integral Equations: A Reference Text, (Noordhoff Int. Publ., Leyden, 1975)

M. L. Krasnov, A. I. Kiselev, and G. I Makarenko, *Problems and Exercises in Integral Equations*, (Mir Publ., Moscow, 1971)
Wolfram Mathematica, http://www.wolfram.com/mathematica/

Chapter 5
Approximate Methods for Solving Integral Equations

5.1 Approximate Solution of the Fredholm Equation by Replacing the Integral by a Finite Sum

Consider the Fredholm equation of the second kind

$$\varphi(x) = \lambda \int_a^b K(x, t)\varphi(t)\mathrm{d}t + f(x). \tag{5.1}$$

The integral can be replaced by a finite sum using any of the approximate integration formulas dividing $[a, b]$ into n equal parts by points $t_0 = a < t_1 < \cdots < t_n < b$.

(1) Rectangles Formula

$$\int_a^b \Phi(t)\mathrm{d}t = \frac{b-a}{n}[\Phi(\tau_1) + \Phi(\tau_2) + \cdots + \Phi(\tau_n)],$$

where τ_i is any point from the interval $\big[t_{i-1}, t_i\big]$, $i = 1, 2, \ldots, n$. The left end of the interval can be taken as a point $\tau_i = t_{i-1}$, or the right end $\tau_i = t_i$ or the middle of the interval $\tau_i = \frac{t_{i-1}+t_i}{2}$.

(2) Trapezium Formula

$$\int_a^b \Phi(t)\mathrm{d}t = \frac{b-a}{2n}\big[\Phi(t_0) + 2(\Phi(t_1) + \cdots + \Phi(t_{n-1})) + \Phi(t_n)\big].$$

© The Author(s), under exclusive license to Springer Nature Singapore Pte Ltd. 2022
V. Ryzhov et al., *Modern Methods in Mathematical Physics*,
https://doi.org/10.1007/978-981-19-4915-9_5

(3) Simpson's 1/3 Formula (or Parabola Formula)

$$\int\limits_a^b \Phi(t)dt = \frac{b-a}{6m}\big[\Phi(t_0) + 2\Phi(2m) + 4(\Phi(t_1) + \cdots + \Phi(t_{2m-1}))$$

$$+2(\Phi(t_2) + \cdots + \Phi(t_{2m-2}))\big],$$

where $n = 2m$.

When using any of these formulas, Eq. (5.1) will be replaced by the following approximate equality:

$$\varphi(x) \approx \lambda \sum_{i=0}^n A_i K(x, t_i)\varphi(t_i) + f(x). \tag{5.2}$$

Setting Eq. (5.2) in succession $x = t_0, t_1, \ldots, t_n$, we obtain a system of linear equations for $\varphi(t_i), i = 0, 1, \ldots, n$. Having solved the system, one can find a solution of the equation $\varphi(x)$, using any interpolation of the obtained discrete values of $\varphi(t_i)$, or simply substituting the found values $\varphi(t_i)$ in (5.2).

More about the use of numerical and approximate methods for solving integral equations can be found in [1–7].

Similarly, one can solve the Fredholm equations of the first kind.

Example 5.1 Solve approximately the equation

$$\varphi(x) = 2 \int\limits_0^1 xe^t \varphi(t)dt + e^{-x}.$$

Solution
Let us write a function in *Wolfram Mathematica* [8] that builds an approximate solution using the trapezoid formula.

```
In [1]:= Clear[NISolveTrapez]
         NISolveTrapez[Kernel_,f_,{a_,b_},φ_,x_,n_]:=
         Block[{kList,xk,Kernelk,yk,fk,KernApprox,IntAp-
prox,
            system,sol,y,t=x},
         kList=Range[0,n];
         xk=a+# (b-a)/n&/@kList;
         Kernelk=Kernel[x,#]&/@xk;
         yk=Subscript[y,#]&/@kList;
         fk=f/@xk;
         KernApprox=Kernelk yk;
```

```
                IntApprox=(b-a)/(2n) (KernApprox[[1]]+KernApp-
rox[[-1]]+
                2If[n>=2,Total@KernApprox[[2;;-2]],0]);
            system=Thread[yk==(λ IntApprox/.x->#&/@xk+fk)];
            sol=Solve[system,yk][[1]]//N;
            φ->Function[Evaluate@t,Evaluate[f[x]+λ IntAp-
prox/.sol]]
        ]
```

Let us set the initial data and use the constructed function, for example, dividing the interval into ten parts.

```
In [3]:= Clear[Kernel,f,a,b,λ,φ,x,n]
         Kernel=Function[{x,t},x Exp[t]];
         f=Function[x,Exp[-x]];
         a=0;
         b=1;
         λ=2;
In[9]:= approxsol=NISolveTrapez[Kernel,f,{a,b},φ,x,10]

Out[9]:= φ → Function[x, e^{-x} − 1.9853243109047263x]
```

It is known that the exact solution $\varphi(x) = -2x + e^{-x}$. Let us set it and build graphs of the exact solution and several approximate solutions for a different number of partitions of the interval (Fig. 5.1).

```
In [10]:= exactsol=-2x+Exp[-x];
```

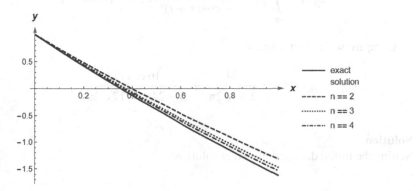

Fig. 5.1 Comparison of exact and approximate solutions of an integral equation obtained by replacing the integral with a finite sum (Example 5.1)

```
          nList={2,3,4};
          approxsols= φ[x]/.NISolveTrapez[Ker-
nel,f,{a,b}, φ,x,#]&/@
          nList;
In[13]:= legend=Prepend[ToString/@Thread[n==#]&@nList,
          "exact solution"];
          optplot={AxesStyle->Arrowheads[{0.0,0.025}],
          AxesLabel->{Style[x,Bold,Me-
dium],Style[y,Bold,Medium],Style[y,Bold,Medium]}};
          Plot[{exactsol,approxsols}//Evaluate,
          {x,a,b},Evaluate@optplot,PlotLegends->legend]
```

To numerically estimate the deviation of approximate solutions for different n from the exact solution, we calculate the norm of their difference.

```
In [16]:= error=Table[{nList[[n]],Sqrt@NIntegrate[(exactsol-
          approxsols[[n]])^2,{x,
-π,π}]},{n,1,Length@nList}];
          error//Column

          {2, 1.4098732814533008}
Out[16]:= {3, 0.6882639668693629}
          {4, 0.4009568123348231}
```

As one can see, with an increase in the number of partitions of an interval, the approximation accuracy is expected to increase.

Example 5.2 Solve approximately the equation

$$\varphi(x) + \int\limits_{-\pi}^{\pi} \varphi(t)\frac{dt}{3 - \cos(x + t)} = 25 - 16\sin^2(x),$$

Compare with exact solution

$$\varphi(x) = \frac{34}{2 + \sqrt{2\pi}} + \frac{16\cos 2x}{2 - 24\pi + 17\sqrt{2\pi}}.$$

Solution
Setting the initial data and the exact solution.

```
In [18]:= Clear[Kernel,f,a,b,λ,φ,x,n]
          Kernel=Function[{x,t},1/(3-Cos[x+t])];
          f=Function[x,25-16Sin[x]^2];
          a=-π;
          b=π;
          λ=-1;
          exactsol=34/(2+Sqrt[2]π)+16Cos[2x]/(2-
24π+17Sqrt[2]π);
```

Let us use the **NISolveTrapez** function from the previous example and plot the graphs of the exact solution and several approximate solutions for a different number of partitions of the interval (Fig. 5.2).

```
In [25]:= nList={2,5,10,20};
          approxsols= φ[x]/.NISolveTrapez[Ker-
nel,f,{a,b},φ,x,#]&/@
          nList;
In[27]:= legend=Prepend[ToString/@Thread[n==#]&@nList,
          "exact solution"];
          optplot={AxesStyle->Arrowheads[{0.0,0.025}],
          AxesLabel->{Style[x,Bold,Medium],Style[y,Bold,Me-
dium],Style[y,Bold,Medium]}};
          Plot[{exactsol,approxsols}//Evaluate,
          {x,a,b},Evaluate@optplot,PlotLegends->legend]
```

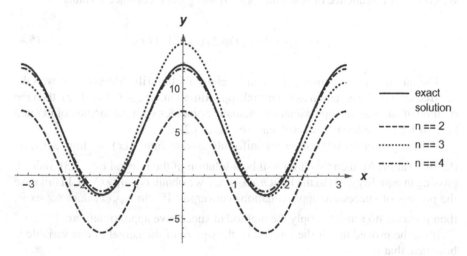

Fig. 5.2 Comparison of exact and approximate solutions of an integral equation obtained by replacing the integral with a finite sum (Example 5.2)

Let us calculate the norm of the difference between the exact and approximate solutions for different n.

```
In[30]:= error=Table[{nList[[n]],Sqrt@NIntegrate[(exactsol-
                        approxsols[[n]])^2,{x,-
π,π}]},{n,1,Length@nList}];
          error//Column

                    {2, 12.123532704730053}
Out[31]:=   {3, 3.748801067774196}
                    {4, 0.7711501673003928}
```

As you can see, with an increase in the number of partitions of a segment, the approximation accuracy is expected to increase.

5.2   Successive Approximation Method

Consider the Fredholm equation of the second kind

$$\varphi(x) = \lambda \int\limits_a^b K(x,t)\varphi(t)\mathrm{d}t + f(x). \tag{5.3}$$

We will solve it by the method of successive approximations (iterations). For this, we construct a sequence of functions $\{\varphi_n(t)\}$ using the recurrence formula

$$\varphi_n(x) = \lambda \int\limits_a^b K(x,t)\varphi_{n-1}(t)\mathrm{d}t + f(x). \tag{5.4}$$

The initial approximation $\varphi_0(x)$ can be chosen arbitrarily. Moreover, if we take the free term of the equation as the initial approximation, i.e., $\varphi_0(x) = f(x)$, then the method of successive approximations actually coincides with the method of iterated kernels for constructing the resolvent (see Sect. 3.2).

If the sequence $\{\varphi_n(t)\}$ converges uniformly to some limit $\varphi(x) = \lim\limits_{n \to +\infty} \varphi_n(x)$ on the interval $[a, b]$, then this limit will be a solution of the integral Eq. (5.3). Indeed, passing in equality (5.4) to the limit $n \to +\infty$, we obtain equality (5.3). In this case the process of successive approximations converges. If $\lim\limits_{n \to +\infty} \varphi_n(x)$ does not exist, then it makes no sense to apply the method of successive approximations.

It can be proved that if the integral of the square of the kernel in one variable is bounded, that is,

$$\int_a^b |K(x,t)|^2 dt < C,$$

then the process of successive approximations converges for all values of the parameter λ, satisfying the inequality $|\lambda| < \frac{1}{B}$, where

$$B = K(x,t) = \sqrt{\int_a^b \int_a^b K^2(x,t) dx dt}.$$

The limit of successive approximations in this case is a solution of the integral equation and this solution is unique. Then, the solution of the equation is approximately equal to the nth approximation $\varphi(x) \approx \varphi_n(x)$, and the error does not exceed the value

$$D\sqrt{C} \frac{|\lambda|^{n+1} B^n}{1-|\lambda|B}, \tag{5.5}$$

where

$$D = f(x) = \sqrt{\int_a^b f^2(x) dx}.$$

Constants B and C are defined above.

The Volterra equations of the second kind are solved similarly by the method of successive approximations. It should be noted that the method of successive approximations for the Volterra equation, in contrast to the Fredholm equation, converges for any values of the parameter λ.

Sometimes, when solving an equation by the method of successive approximations, it is possible to determine the general pattern of expression of iterations $\varphi_n(x)$ and calculate $\lim_{n \to +\infty} \varphi_n(x)$, that is, find the exact solution to the equation.

The Fredholm and Volterra equations of the first kind can also be solved by the method of successive approximations, having previously transformed them to the standard form

$$\varphi(x) = F(x, \varphi(x)).$$

For example, the Fredholm equation of the first kind

$$\int_a^b K(x,t)\varphi(t) dt = f(x) \tag{5.6}$$

can be written as

$$\varphi(x) = f(x) + \left[\varphi(x) - \int\limits_a^b K(x,t)\varphi(t)\mathrm{d}t \right],\tag{5.7}$$

and then each subsequent approximation is found through the previous one by the formula

$$\varphi_{n+1}(x) = f(x) + \left[\varphi_n(x) - \int\limits_a^b K(x,t)\varphi_n(t)\mathrm{d}t \right], n = 0, 1, \ldots;$$

$$\varphi_0(x) = f(x).$$

It is useful to note that writing the equation of the first kind (5.6) in the form (5.7) is not the only possible way. Equation (5.6) can also be written as

$$\varphi(x) = f(x) - \left[\int\limits_a^b K(x,t)\varphi(t)\mathrm{d}t - \varphi(x) \right].\tag{5.8}$$

The convergence of the method of successive approximations may depend on the choice of the form of writing the equation.

Along with formulas (5.7) and (5.8), the following notation is also used

$$\varphi(x) = \varphi(x) + \lambda \left[f(x) - \int\limits_a^b K(x,t)\varphi(t)\mathrm{d}t \right].\tag{5.9}$$

The parameter λ is selected so that the iteration process converges.

The method of successive approximations can also be used to construct approximate solutions for nonlinear integral equations. For example, for the Volterra nonlinear integral equation

$$\varphi(x) = f(x) + \int\limits_a^x F(x,t,\varphi(t))\mathrm{d}t\tag{5.10}$$

successive approximations $\varphi_n(x)$ can be calculated by the formula

$$\varphi_n(x) = f(x) + \int\limits_a^x F(x,t,\varphi_{n-1}(t))\mathrm{d}t,\tag{5.11}$$

where $\varphi_0(x)$ can be chosen arbitrarily.

Example 5.3 Solve the Fredholm equation by the method of successive approximations

$$\varphi(x) = 0.1 \int\limits_0^1 K(x,t)\varphi(t)\mathrm{d}t + 1,$$

$$K(x,t) = \begin{cases} x, \ 0 \le x \le t \\ t, \ t \le x \le 1 \end{cases}.$$

Solution
Set the initial data.

```
In [1]:= Clear[Kernel,f,a,b,λ,φ,x,n]
         Kernel=Function[{x,t},Piece-
wise[{{x,a<=x<=t},{t,t<=x<=b}}]];
         f=Function[x,1];
         a=0;
         b=1;
         λ=0.1;
```

We calculate the kernel norm, the constant B and check the inequality $|\lambda| < \frac{1}{B}$.

```
In [7]:= B=Sqrt@Integrate[Kernel[x,t]^2,{x,a,b},{t,a,b}]
         Abs@λ<1/B
```

Out[7] := $\frac{1}{\sqrt{6}}$
Out[8] := **True**

Let us define a function for constructing a sequence of approximations. As an initial approximation, we take the free term of the equation, i.e., $\varphi_0(x) = 1$.

```
In [9]:= Clear[φnFred]
         φnFred[Kernel_,f_,λ_,{a_,b_},x_,φ0_,1]:=
          λ Integrate[Kernel[x,t]φ0[t],{t,a,b},
            Assumptions->0<x<1]+f[x]//Expand
         φnFred[Kernel_,f_,λ_,{a_,b_},x_,φ0_,n_/;n>=2]:=Mod-
ule[{t},
          λ Integrate[Kernel[x,t]φnFred[Ker-
nel,f,λ,{a,b},t,φ0,n-1],
            {t,a,b},Assumptions->0<x<1]+f[x]//Expand
          ]
```

Construct approximations for n from 1 to 5.

```
In [12]:= nList=Range[5];
          φnList=φnFred[Kernel,f,λ,{a,b},x,f,#]&/@nList;
          φ#==ScientificForm[φnList[[#]],3]&/@
          Range[Length@nList]//Column
```

$$Out[14] := \phi_1 == 1 + \left(1. \times 10^{-1}\right)x - \left(5. \times 10^{-2}\right)x^2$$
$$\phi_2 == 1 + \left(1.03 \times 10^{-1}\right)x - \left(5. \times 10^{-2}\right)x^2 -$$
$$\left(1.67 \times 10^{-3}\right)x^3 + \left(4.17 \times 10^{-4}\right)x^4$$
$$\phi_3 == 1 + \left(1.03 \times 10^{-1}\right)x - \left(5. \times 10^{-2}\right)x^2 -$$
$$\left(1.72 \times 10^{-3}\right)x^3 + \left(4.17 \times 10^{-4}\right)x^4 +$$
$$\left(8.33 \times 10^{-6}\right)x^5 - \left(1.39 \times 10^{-6}\right)x^6$$
$$\phi_4 == 1 + \left(1.03 \times 10^{-1}\right)x - \left(5. \times 10^{-2}\right)x^2 -$$
$$\left(1.72 \times 10^{-3}\right)x^3 + \left(4.17 \times 10^{-4}\right)x^4 +$$
$$\left(8.61 \times 10^{-6}\right)x^5 - \left(1.39 \times 10^{-6}\right)x^6 -$$
$$\left(1.98 \times 10^{-8}\right)x^7 + \left(2.48 \times 10^{-9}\right)x^8$$
$$\phi_5 == 1 + \left(1.03 \times 10^{-1}\right)x - \left(5. \times 10^{-2}\right)x^2 -$$
$$\left(1.72 \times 10^{-3}\right)x^3 + \left(4.17 \times 10^{-4}\right)x^4 +$$
$$\left(8.62 \times 10^{-6}\right)x^5 - \left(1.39 \times 10^{-6}\right)x^6 -$$
$$\left(2.05 \times 10^{-8}\right)x^7 + \left(2.48 \times 10^{-9}\right)x^8 +$$
$$\left(2.76 \times 10^{-11}\right)x^9 - \left(2.76 \times 10^{-12}\right)x^{10}$$

Now, estimate the errors for the constructed approximations by the formula (5.5).

```
In [15]:= Normf=Sqrt@Integrate[1,{t,a,b}];
          c=Maximize[{Integrate[Kernel[x,t]^2,{t,a,b},
              Assumptions->0<=x<=1],0<=x<=1},x][[1]];
          errorList=Normf Sqrt@1 Abs[λ]^(#+1)NormK^#/
              (1-Abs@λ NormK)&/@nList;
          ScientificForm[errorList, 3]
```

$$Out[18] := \left\{4.26 \times 10^{-3}, 1.74 \times 10^{-4}, 7.09 \times 10^{-6}, 2.9 \times 10^{-7}, 1.18 \times 10^{-8}\right\}$$

We will also plot the graphs of some of the obtained approximations (Fig. 5.3).

Fig. 5.3 Approximate solutions of an integral equation by the method of successive approximations, segment [0, 1] (Example 5.3)

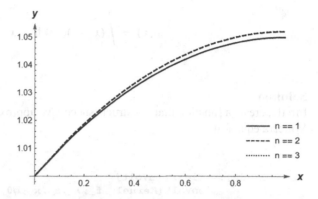

```
In [19]:= legend=ToString/@Thread[n==#]&@nList[[1;;3]];
          optplot={AxesStyle->Arrowheads[{0.0,0.025}],
          AxesLabel->{Style[x,Bold,Medium],Style[y,Bold,Me-
dium],Style[y,Bold,Medium]}};
          Plot[φnList[[1;;3]],{x,a,b},
          Evaluate@optplot,PlotLegends->legend]
```

We can look at the segment [0.8, 1] on a larger scale to see how close the approximations are (Fig. 5.4).

According to the error estimates and plots, it can be seen that already starting with $n = 2$, we get a fairly good approximation of the solution to the given integral equation.

Example 5.4 Solve the Volterra equation by the method of successive approximations.

Fig. 5.4 Approximate solutions of an integral equation by the method of successive approximations, segment [0.8, 1] (Example 5.3)

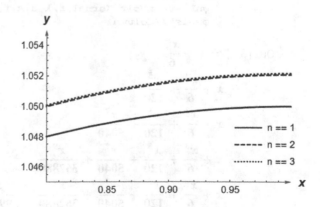

$$\varphi(x) = \int\limits_0^x (x-t)\varphi(t)\mathrm{d}t + x.$$

Solution

Firstly, create a function that constructs successive approximations in the case of the Volterra equation.

```
In [22]:= Clear[φnVolt]
          φnVolt[Kernel_,f_,λ_,a_,x_,φ0_,1]:=
            λ Integrate[Kernel[x,t]φ0[t],{t,a,x},
               Assumptions->0<x<1]+f[x]//Expand
          φnVolt[Kernel_,f_,λ_,a_,x_,φ0_,n_/;n>=2]:=Mod-
ule[{t},
              λ Integrate[Kernel[x,t]φnVolt[Ker-
nel,f,λ,a,t,φ0,n-1],
                 {t,a,x},Assumptions->0<x<1]+f[x]//Expand
            ]
```

Secondly, set the initial data and calculate several approximations. For the initial approximation, we take the free term of the equation, i.e., $\varphi_0(x) = x$.

```
In [25]:= Clear[Kernel,f,a,b,λ,φ,x,n]
          Kernel=Function[{x,t},x-t];
          f=Function[x,x];
          a=0;
          λ=1;
In[30]:= nList=Range[5];
          φnList=φnVolt[Kernel,f,λ,a,x,f,#]&/@nList;
          φnList//Column
```

$$\text{Out[32]} := x + \frac{x^3}{6}$$

$$x + \frac{x^3}{6} + \frac{x^5}{120}$$

$$x + \frac{x^3}{6} + \frac{x^5}{120} + \frac{x^7}{5040}$$

$$x + \frac{x^3}{6} + \frac{x^5}{120} + \frac{x^7}{5040} + \frac{x^9}{362880}$$

$$x + \frac{x^3}{6} + \frac{x^5}{120} + \frac{x^7}{5040} + \frac{x^9}{362880} + \frac{x^{11}}{39916800}$$

Thirdly, determine the pattern and sum up the functional series. To do this, we extract the coefficients and obtain a general formula for them. In this case, we will take into account that the coefficients at even powers of x vanish.

```
In[33]:= coefs=DeleteCases[CoefficientList[φnVolt[Kernel,
              f,λ,a,x,f,5],x],0]
```

$$\text{Out}[33] := \left\{1, \frac{1}{6}, \frac{1}{120}, \frac{1}{5040}, \frac{1}{362880}, \frac{1}{39916800}\right\}$$

```
In[34]:= summand=FullSimplify[FindSequenceFunction[co-
efs,n]x2n-1]/.
              Gamma[2 n]->(2n-1)!
```

$$\text{Out}[34] := \frac{x^{-1+2n}}{\text{Gamma}[2n]}$$

Next, calculate the sum of the series. For this, for example, the **AsymptoticSum** function is suitable.

```
In[35]:= sol=AsymptoticSum[summand,{n,1,k},k->∞]
```

$$\text{Out}[35] := \mathbf{Sinh}[x]$$

Finally, we check the solution by substituting it into the equation.

```
In[36]:= φ[x]==λ Integrate[Kernel[x,t]φ[t],{t,a,x}]+f[x]/.
              φ->Function[x,Evaluate@sol]
```

$$\text{Out}[36] := \mathbf{True}$$

Thus, we have obtained an exact solution to the equation using the method of successive approximations. However, we note that the exact solution cannot always be obtained.

Example 5.5 Solve the Fredholm equation of the first kind using the method of successive approximations,

$$\int_0^1 K(x,t)\varphi(t)\mathrm{d}t = \sin(x),$$

where

$$K(x, t) = \begin{cases} (1 - x)t, \, 0 \leq t \leq x \\ (1 - t)x, \, x \leq t \leq 1 \end{cases}.$$

Solution

Since the equation is of the first kind, we will use formula (5.7). Let us create a function to implement it.

```
In [37]:= Clear[ϕnFred2Kind]
          ϕnFred2Kind[Ker-
nel_,f_,{a_,b_},x_,ϕ0_,1]:=f[x]+ϕ0[x]-
              Integrate[Kernel[x,t]ϕ0[t],{t,a,b},
              Assumptions->a<x<b]//Expand
          ϕnFred2Kind[Ker
nel_,f_,{a_,b_},x_,ϕ0_,n_/;n>=2]:=Module[{t},
              f[x]+ϕnFred2Kind[Kernel,f,{a,b},x,ϕ0,n-1]-
              Integrate[Kernel[x,t]
                 ϕnFred2Kind[Kernel,f,{a,b},t,ϕ0,n-
-1],{t,a,b},
              Assumptions->a<x<b]//Expand
          ]
```

Set the initial data and calculate several approximations. For the initial approximation, we take the free term of the equation, i.e., $\varphi_0(x) = \sin x$.

```
In [40]:= Clear[Kernel,f,a,b,ϕ,x,n]
          Kernel=Function[{x,t},
            Piecewise[{{(1-x)t,a<=t<=x},{(1-t)x,x<=t<=b}}]];
          f=Function[x,Sin[π x]];
          a=0;
          b=1;
In[45]:= nList=Range[5];
          ϕnList=ϕnFred2Kind[Kernel,f,λ,a,x,f,#]&/@nList;
          ϕnList//Column
```

$$\text{Out}[47] := 2\text{Sin}[\pi x] - \frac{\text{Sin}[\pi x]}{\pi^2}$$

$$3\text{Sin}[\pi x] + \frac{\text{Sin}[\pi x]}{\pi^4} - \frac{3\text{Sin}[\pi x]}{\pi^2}$$

$$4\text{Sin}[\pi x] - \frac{\text{Sin}[\pi x]}{\pi^6} + \frac{4\text{Sin}[\pi x]}{\pi^4} - \frac{6\text{Sin}[\pi x]}{\pi^2}$$

$$5\text{Sin}[\pi x] + \frac{\text{Sin}[\pi x]}{\pi^8} - \frac{5\text{Sin}[\pi x]}{\pi^6} + \frac{10\text{Sin}[\pi x]}{\pi^4} - \frac{10\text{Sin}[\pi x]}{\pi^2}$$

$$6\text{Sin}[\pi x] - \frac{\text{Sin}[\pi x]}{\pi^{10}} + \frac{6\text{Sin}[\pi x]}{\pi^8} - \frac{15\text{Sin}[\pi x]}{\pi^6} + \frac{20\text{Sin}[\pi x]}{\pi^4} - \frac{15\text{Sin}[\pi x]}{\pi^2}$$

Let us select the law for the sequence and get the exact solution.

```
In [48]:= seqn=FindSequenceFunction[φnList,n]
          sol=DiscreteAsymptotic[seqn,n->∞]
```

$$\mathrm{Out[48]} := -\left(\left(-\left(1 - \frac{1}{\pi^2}\right)^n - \pi^2 + \left(1 - \frac{1}{\pi^2}\right)^n \pi^2\right)\mathrm{Sin}[\pi x]\right)$$

$$\mathrm{Out[49]} := \pi^2\mathrm{Sin}[\pi x]$$

Let us check the obtained solution by substituting it into the original equation.

```
In [50]:= Integrate[Kernel[x,t]φ[t],{t,a,b},Assumptions-
>a<x<b]==
                f[x]/.φ->Function[x,Evaluate@sol]//FullSimplify
```

$\mathrm{Out[50]} := $ **True**

We again managed to obtain an exact solution.

Example 5.6 Find a solution to the Fredholm integral equation using the method of successive approximations

$$\varphi(x) = 1 + \int\limits_0^1 xt^2\varphi(t)\mathrm{d}t.$$

Solution
Set the initial data.

```
In [51]:= Clear[Kernel,f,a,b,λ,φ,x,n]
          Kernel=Function[{x,t},x t^2];
          f=Function[x,1];
          a=0;
          b=1;
          λ=1;
```

Calculate ten approximations for different initial functions: $\varphi_0(x) = f(x) = 1$, $\varphi_0(x) = x$, $\varphi_0(x) = \sin x$.

```
In [57]:= φnFred[Kernel,f,λ,{a,b},x,Function[x,1],10]//N
         φnFred[Kernel,f,λ,{a,b},x,Function[x,x],10]//N
         φnFred[Kernel,f,λ,{a,b},x,Func-
tion[x,Sin[x]],10]//N
```

$$\text{Out}[57] := 1. + 0.4444440205891927x$$
$$\text{Out}[58] := 1. + 0.4444437026977539x$$
$$\text{Out}[59] := 1. + 0.4444436006327648x$$

Let us compare the obtained approximations with the exact answer $\varphi(x) = 1 + \frac{4}{9}x$.

```
In [60]:= 1+4/9x//N
```

$$\text{Out}[60] := 1. + 0.444444444444444x$$

Thus, we see that successive approximations converge to the solution regardless of the initial approximation. However, the error depends on the initial approximation.

Example 5.7 Find a solution of the Volterra nonlinear integral equation using the method of successive approximations

$$\varphi(x) = \int\limits_0^x \frac{1 + \varphi^2(t)}{1 + t^2} dt.$$

Solution
The previously constructed function for the linear Volterra equations will not work for this case. So, we will construct a new one.

```
In [61]:= Clear[φnVoltNonLinear]
          φnVoltNonLinear[Kernel_,f_,a_,φ0_,1]:=
          Function[var,Evaluate[f[var]+
              Integrate[Kernel[var,t,φ0],{t,a,var},
                GenerateConditions->False]//Expand]]
          φnVoltNonLinear[Kernel_,f_,a_,φ0_,n_/;n>=2]:=
          Function[var,Evaluate[f[var]+
              Integrate[Kernel[var,t,
                φnVoltNonLinear[Kernel,f,a,φ0,n-
```

```
1]],{t,a,var},
                    GenerateConditions->False]//Expand]]
```

Set the initial data. Now, the **Kernel** will also depend on the desired function.

```
In [64]:= Clear[Kernel,f,a,b,λ,ϕ,x,n]
          Kernel=Function[{x,t,ϕ},(1+ϕ[t]^2)/(1+t^2)];
          f=Function[x,0];
          a=0;
          eqn=ϕ[x]==f[x]+Integrate[Kernel[x,t,ϕ],{t,a,x}]
```

$$\text{Out[68]}:= \phi[x] == \int_0^x \frac{1+\phi[t]^2}{1+t^2}\,dt$$

Take $\varphi_0(x) = 0$ as an initial approximation and calculate several iterations.

```
In [69]:= nList=Range[5];
          ϕnList=ϕnVoltNonLinear[Kernel,f,a,f,#][x]&/@nList;
          ϕnList[[1;;4]]//Column
```

$$\text{Out[71]}:= \text{ArcTan}[x]$$

$$\text{ArcTan}[x] + \frac{\text{ArcTan}[x]^3}{3}$$

$$\text{ArcTan}[x] + \frac{\text{ArcTan}[x]^3}{3} + \frac{2\,\text{ArcTan}[x]^5}{15} + \frac{\text{ArcTan}[x]^7}{63}$$

$$\text{ArcTan}[x] + \frac{\text{ArcTan}[x]^3}{3} + \frac{2\,\text{ArcTan}[x]^5}{15} + \frac{17\,\text{ArcTan}[x]^7}{315} +$$

$$\frac{38\,\text{ArcTan}[x]^9}{2835} + \frac{134\,\text{ArcTan}[x]^{11}}{51975} + \frac{4\,\text{ArcTan}[x]^{13}}{12285} + \frac{\text{ArcTan}[x]^{15}}{59535}$$

It seems as if the coefficients at the arctangent form a sequence. Let us try to get it.

```
In [72]:= FindSequenceFunction[DeleteCases[
          CoefficientList[ϕnList[[-1]],ArcTan[x]],0],n]
```

$$\mathrm{Out[72]} := \mathrm{FindSequenceFunction}\Big[\Big\{1, \frac{1}{3}, \frac{2}{15}, \frac{17}{315}, \frac{62}{2835}, \frac{1142}{155925}, \frac{13324}{6081075},$$
$$\frac{377017}{638512875}, \frac{1522814}{10854718875}, \frac{24022}{820945125}, \frac{29756}{5746615875}, \frac{12676238}{16962094524375},$$
$$\frac{256948}{3016973334375}, \frac{100732}{14119435204875}, \frac{8}{21210236775}, \frac{1}{109876902975}\Big\}, n\Big]$$

However, it was not possible to find a law for this sequence. Then, we will build graphs of approximations (Fig. 5.5).

```
In [73]:= legend=ToString/@Thread[n==#]&@nList;
          optplot={AxesStyle->Arrowheads[{0.0,0.025}],
          AxesLabel->{Style[x,Bold,Medium],Style[y,Bold,Me-
dium],Style[y,Bold,Medium]}};
          Plot[ϕnList,{x,a-5,a+5},Evaluate@optplot,
          PlotLegends->legend]
```

Let us try again to build approximations by taking $\varphi_0(x) = x$.

```
In [76]:= nList=Range[5];
          ϕnList=ϕnVoltNonLinear[Kernel,f,a,Func-
tion[x,x],#][x]&/@
          nList;
          ϕnList[[1;;4]]//Column
```

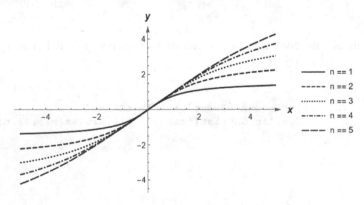

Fig. 5.5 Approximate solutions of an integral equation by the method of successive approximations (Example 5.7)

$$\text{Out[78]} := \begin{matrix} x \\ x \\ x \\ x \end{matrix}$$

This means that the exact solution is $\varphi(x) = x$.

Example 5.8 Find a solution of the Volterra nonlinear integral equation using the method of successive approximations

$$\varphi(x) = 1 + \int_0^x \left[\varphi^2(t) + t\varphi(t) + t^2 \right] dt.$$

Solution

Let us set the initial data and use the function from the previous example, taking $\varphi_0(x) = f(x) = 1$.

```
In [79] := Clear[Kernel,f,a,b,λ,ϕ,x,n]
           Kernel=Function[{x,t,ϕ},ϕ[t]^2+t ϕ[t]+t^2];
           f=Function[x,1];
           a=0;
In[83] := nList=Range[5];
          ϕnList=ϕnVoltNonLinear[Kernel,f,a,f,#][x]&/@nList;
          ϕnList[[1;;3]]//Column
```

$$\text{Out[85]} := 1 + x + \frac{x^2}{2} + \frac{x^3}{3}$$

$$1 + x + \frac{3x^2}{2} + \frac{4x^3}{3} + \frac{13x^4}{24} + \frac{x^5}{4} + \frac{x^6}{18} + \frac{x^7}{63}$$

$$1 + x + \frac{3x^2}{2} + 2x^3 + \frac{43x^4}{24} + \frac{22x^5}{15} + \frac{49x^6}{48} + \frac{307x^7}{504} + \frac{67x^8}{224} + \frac{4735x^9}{36288} +$$

$$\frac{1411x^{10}}{30240} + \frac{499x^{11}}{33264} + \frac{17x^{12}}{4536} + \frac{25x^{13}}{29484} + \frac{x^{14}}{7938} + \frac{x^{15}}{59535}$$

Plot the graph of the obtained approximations (Fig. 5.6).

```
In [86] := legend=ToString/@Thread[n==#]&@nList;
           optplot={AxesStyle->Arrowheads[{0.0,0.025}],
               AxesLabel->{Style[x,Bold,Me-
dium],Style[y,Bold,Medium],Style[y,Bold,Medium]},
```

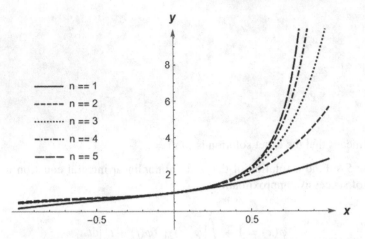

Fig. 5.6 Approximate solutions of an integral equation by the method of successive approximations, $\varphi_0(x) = 1$ (Example 5.8)

```
      PlotRange->{-1,10}}};
      Plot[φnList,{x,a-1,a+1},Evaluate@optplot,
      PlotLegends->legend]
```

The graph shows that process converges. Now, let us try to take $\varphi_0(x) = 1 + x$ and plot the graphs (Fig. 5.7).

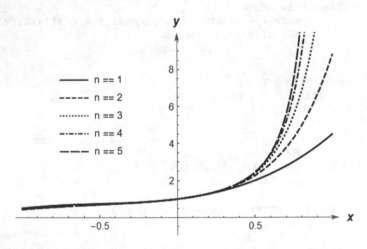

Fig. 5.7 Approximate solutions of the integral equation by the method of successive approximations, $\varphi_0(x) = 1 + x$ (Example 5.8)

```
In [89]:= nList=Range[5];
          ϕnList=ϕnVoltNonLinear[Kernel,f,a,Func-
tion[x,1+x],#][x]&/@
          nList;
          ϕnList[[1;;3]]//Column
```

$$Out[91] := 1 + x + \frac{3x^2}{2} + x^3$$

$$1 + x + \frac{3x^2}{2} + 2x^3 + \frac{13x^4}{8} + \frac{21x^5}{20} + \frac{x^6}{2} + \frac{x^7}{7}$$

$$1 + x + \frac{3x^2}{2} + 2x^3 + \frac{17x^4}{8} + \frac{23x^5}{10} + \frac{173x^6}{80} + \frac{521x^7}{280} + \frac{1601x^8}{1120} +$$

$$\frac{19643x^9}{20160} + \frac{3271x^{10}}{5600} + \frac{9237x^{11}}{30800} + \frac{53x^{12}}{420} + \frac{11x^{13}}{260} + \frac{x^{14}}{98} + \frac{x^{15}}{735}$$

```
In [92]:= legend=ToString/@Thread[n==#]&@nList;
          optplot={AxesStyle->Arrowheads[{0.0,0.025}],
          AxesLabel->{Style[x,Bold,Medium],Style[y,Bold,Me-
dium],Style[y,Bold,Medium]},
              PlotRange->{-1,10}};
          Plot[ϕnList,{x,a-1,a+1},Evaluate@optplot,
          PlotLegends->legend]
```

We can see that this initial approximation turns out to be slightly better.

5.3 Bubnov–Galerkin Method

An approximate solution of the Fredholm equation of the second kind

$$\varphi(x) = f(x) + \lambda \int_a^b K(x,t)\varphi(t)\mathrm{d}t \tag{5.12}$$

can be found by the Bubnov–Galerkin method as follows. We choose a system of functions $\{u_n(x)\}$, complete in $L_2(a,b)$ and such that for any n the functions $u_0(x), u_1(x), \ldots, u_n(x)$ are linearly independent, and we look for an approximate solution $\varphi_n(x)$ as an expansion in the system

$$\varphi_n(x) = \sum_{k=0}^{n} a_k u_k(x). \tag{5.13}$$

The coefficients a_k can be found by solving the following linear system:

$$(\varphi_n(x), u_k(x)) = (f(x), u_k(x)) + \lambda\left(\int\limits_a^b K(x, t)\varphi_n(t)\mathrm{d}t, u_k(x)\right). \qquad (5.14)$$

Here, the scalar product is understood as the integral

$$(f(x), g(x)) = \int\limits_a^b f(x)g(x)\mathrm{d}x.$$

The expression (5.13) is substituted into the formula (5.14) for $\varphi_n(x)$. If the value of the parameter λ is not characteristic, then for sufficiently large values of n, the constructed approximate solution $\varphi_n(x)$ tends to the solution $\varphi(x)$ in the metric of functions summable with the square in $L_2(a, b)$ space.

As a system of functions complete on the interval $[-1, 1]$ the Legendre polynomials $P_n(x)$ or the Chebyshev polynomials $T_n(x)$ can be used.

Example 5.9 Find a solution of the Fredholm integral equation using the Bubnov–Galerkin method

$$\varphi(x) = x + \int\limits_{-1}^1 xt\varphi(t)\mathrm{d}t.$$

Solution

Let us define the **BubGal** function for building solutions. We will use the Legendre polynomials $P_n(x)$ or the Chebyshev polynomials $T_n(x)$ as the complete system of functions. In this case, the choice of polynomials will be implemented through the *functions* option.

```
In [1]:= Clear[BubGal]
         Options[BubGal]={functions->"Legen-
dre"};
         BubGal[Kernel_,f_,n_,ϕ_,var_,OptionsPattern[]]:=
         Block[{cList,c,uList,ϕn,cSys,cSol,t,x},
         cList=Subscript[c,#]&/@Range[0,n];
         uList[x_]:=Switch[OptionValue[functions],"Legen-
dre",
             LegendreP[#,x]&/@Range[0,n],"Chebyshev",
             ChebyshevT[#,x]&/@Range[0,n]];
         ϕn[x_]=cList.uList[x];
         cSys=Table[Integrate[ϕn[x] uList[x][[k+1]],{x,-
1,1}]==
             Integrate[f[x] uList[x][[k+1]],{x,-1,1}]+
             Integrate[Kernel[x,t]  ϕn[t]uList[x][[k+1]],
             {t,-1,1},{x,-1,1}],{k,0,n} ];
```

```
cSol=Solve[cSys,cList][[1]];
ϕ->Function[x,ϕn[x]/.cSol//Evaluate]/.x->var
]
```

Set the initial data and construct a solution for n from 0 to 5, using the Legendre polynomials.

```
In [4]:= Clear[Kernel,f,a,b,ϕ,x,n]
         Kernel=Function[{x,t},x t];
         f=Function[x,x];
         a=-1;
         b=1;
         eqn=ϕ[x]==f[x]+Integrate[Kernel[x,t]ϕ[t],{t,a,b}]
```

$$Out[9] := \phi[x] == x + \int_{-1}^{1} tx\phi[t]\mathrm{d}t$$

```
In [10]:= ϕ[x]/.BubGal[Kernel,f,#,ϕ,x,functions->"Legen-
dre"]&/@
          Range[0,5]
```

$$Out[10] := \{0, 3x, 3x, 3x, 3x, 3x\}$$

We see that starting from $n = 1$ the solution does not change. This means that most likely we have received an exact solution. The following code proves that we obtain the exact solution.

```
In [11]:= eqn/.ϕ->Function[x,3 x]
```

$Out[11] := $ **True**

We will also construct a solution using the Chebyshev polynomials.

```
In [12]:= ϕ[x]/.BubGal[Kernel,f,#,ϕ,x,functions->"Cheby-
shev"]&/@
          Range[0,5]
```

$$Out[12] := \{0, 3x, 3x, 3x, 3x, 3x\}$$

We also obtain an exact solution for $n = 1$.

Thus, the Bubnov–Galerkin method made it possible to find the exact solution of this equation: $\varphi(x) = 3x$.

Example 5.10 Find a solution of the Fredholm integral equation using the Bubnov–Galerkin method

$$\varphi(x) = 1 + \int_{-1}^{1} (xt + x^2)\varphi(t)dt.$$

Solution
Set the initial data and use the **BubGal** function from the previous example. Let us try to build a solution in the form of an expansion in terms of Legendre and Chebyshev polynomials.

```
In [13]:= Clear[Kernel,f,a,b,ϕ,x,n]
          Kernel=Function[{x,t},x t+x^2];
          f=Function[x,1];
          a=-1;
          b=1;
          eqn=ϕ[x]==f[x]+Integrate[Kernel[x,t]ϕ[t],{t,a,b}]
```

$$Out[18] := \phi[x] == 1 + \int_{-1}^{1} (tx + x^2)\phi[t]dt$$

```
In [19]:= ϕ[x]/.BubGal[Kernel,f,#,ϕ,x,functions->"Legen-
          dre"]&/@
          Range[0,5]
```

$$Out[19] := \begin{array}{l} 3 \\ 3 \\ 3+2(-1+3x^2) \\ 3+2(-1+3x^2) \\ 3+2(-1+3x^2) \\ 3+2(-1+3x^2) \end{array}$$

```
In [20]:= ϕ[x]/.BubGal[Kernel,f,#,ϕ,x,functions->"Cheby-
          shev"]&/@
          Range[0,5]
```

$$Out[20] := \begin{array}{l} 3 \\ 3 \\ 4+3(-1+2x^2) \\ 4+3(-1+2x^2) \\ 4+3(-1+2x^2) \\ 4+3(-1+2x^2) \end{array}$$

We can see that starting from $n = 2$ the solution is stabilized and does not depend on the expansion in polynomials: $\varphi(x) = 6x^2 + 1$.

```
In [21]:= eqn/.φ->Function[x,6x^2+1]

Out[21]:= True
```

5.3.1 Method of Replacing a Kernel with a Degenerate One

Consider the Fredholm integral equation of the second kind

$$\varphi(x) = f(x) + \lambda \int_a^b K(x,t)\varphi(t)dt. \tag{5.15}$$

with an arbitrary kernel $K(x,t)$. We replace this kernel with an approximate degenerate

$$L(x,t) = \sum_{k=1}^n a_k(x)b_k(t).$$

As a result, we get a new equation

$$z(x) = f(x) + \lambda \int_a^b L(x,t)z(t)dt, \tag{5.16}$$

where $z(x)$ is taken as an approximation to the solution $\varphi(x)$ of the original equation. The methods for solving equations with degenerate kernels can be applied to Eq. (5.16): reduction to an algebraic equation or the method of resolvents (see Chap. 3).

Methods for constructing degenerate kernels $L(x,t)$, close to a given nondegenerate kernel $K(x,t)$, are different. For example, you can approximate the kernel $K(x,t)$ by partial sums of power series or double Fourier series converging uniformly to the kernel $K(x,t)$ in the square $[a,b] \times [a,b]$, or approximate it by algebraic or trigonometric interpolation polynomials.

It can be shown that if the sequence $\{L_n(x,t)\}$ of degenerate kernels converges uniformly to the kernel $K(x,t)$, then the sequence of solutions $\{z_n(x)\}$ of the equation with kernels $L_n(x,t)$ and the same free term $f(x)$ will uniformly converge to the solution $\varphi(x)$ of the equation with kernel $K(x,t)$.

Example 5.11 Approximately, solve the Fredholm equation of the second kind, replacing the kernel of the equation by the degenerate one

$$\varphi(x) = \int\limits_0^1 x\left(1 - e^{xt}\right)\varphi(t)dt + e^x - x.$$

Solution

We define a function for approximating a nondegenerate kernel by the sum of the first terms of the Taylor series expansion in powers of t, i.e.,

$$L(x, t) = \sum_{k=0}^n a_k(x)t^k.$$

After replacing the kernel with a degenerate one, the integral equation is reduced to a system of algebraic equations.

```
In [1]:= Clear[ApproxKernelByTaylor]
         ApproxKernelByTeylor[Kernel_,f_,{a_,b_},ϕ_,x_,n_]:=
         Block[{L,funcList,aList,bList,cList,m,c,coef-
Matrix,
            fList,cSol,s=x,t},
          L=Normal@Series[Kernel[s,t],{t,0,n}];
          funcList=If[MatchQ[L,Plus[_,__]],List@@L,{L}];
          aList=funcList/.funcs_ Longest[nots__]:>
              funcs/;FreeQ[{nots},s];
          bList=funcList/aList;
          m=Length@aList;
          cList=Subscript[c,#]&/@Range@m;
          gensol=cList.aList+f[s];
          coefMatrix=Integrate[Outer[Times,bList,aList]/.s-
          coefMatrix=Integrate[Outer[Times,bList,aList]/.s-
>t,
            {t,a,b}];
          fList=Integrate[f[t]*#,{t,a,b}]&/@bList;
          cSol=Solve[Thread[(IdentityMatrix[m]-
              λ coef-
Matrix].cList==fList],cList][[1]]//N;
          ϕ->Function[Evalu
ate@s,Evaluate[cList.aList+f[s]/.cSol]]
         ]
```

Set the data and build a solution by approximating the kernel, for example, to the degree t^n.

```
In [3]:= Clear[Kernel,f,a,b,λ,ϕ,x,n]
         Kernel=Function[{x,t},x(1-Exp[x t])];
```

```
f=Function[x,Exp[x]-x];
a=0;
b=1;
λ=1;
eqn=φ[x]==f[x]+Integrate[Kernel[x,t]φ[t],{t,a,b}];
ApproxKernelByTaylor[Kernel,f,{a,b},φ,x,3]
```

$$\text{Out}[10] := e^x - x - 0.5010191603083365x^2 - 0.1671257525365753x^3 - \\ -0.04180539464358569x^4$$

For a given equation $\varphi(x) = 1$. Let us set it and check it by substitution.

```
In [11]:= exactsol=1;
          eqn/.φ->Function[x,Evaluate@exactsol]
```

$$\text{Out}[12] := \textbf{True}$$

Let us compare graphically the exact solution and approximate solutions for different n (Fig. 5.8).

```
In [13]:= nList=Range[3];
          approxsols=φ[x]/.ApproxKernelByTaylor[Kernel,
             f,{a,b},φ,x,#]&/@nList;
          legend=Append[ToString/@Thread[n==#]&@nList,
            "exact solution"];
```

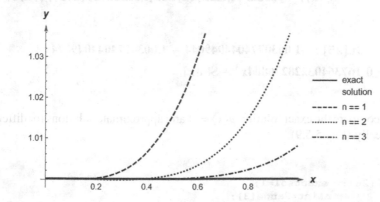

Fig. 5.8 Exact and approximate solutions of an integral equation obtained by replacing the kernel with a degenerate one (Example 5.11)

```
        optplot={AxesStyle->Arrowheads[{0.0,0.025}],
              AxesLabel->{Style[x,Bold,Me-
dium],Style[y,Bold,Medium],Style[y,Bold,Medium]},
           PlotRange->Automatic};
        Plot[{approxsols,exactsol}//Evaluate,{x,a,b},
        Evaluate@optplot,PlotLegends->legend]
```

It can be seen that with increasing n the accuracy of the approximation of the solution also increases.

Example 5.12 Approximately solve the Fredholm equation of the second kind, replacing the kernel of the equation by the degenerate one

$$\varphi(x) = \sin x + \int_0^1 (1 - x \cos xt)\varphi(t)\mathrm{d}t.$$

Solution
Set the initial data and use the **ApproxKernelByTaylor** function from the previous example. We construct an approximate solution by approximating the kernel, for example, by a sum up to t^3.

```
In [18]:= Clear[Kernel,f,a,b,λ,ϕ,x,n]
        Kernel=Function[{x,t},1-x Cos[x t]];
        f=Function[x,Sin[x]];
        a=0;
        b=1;
        λ=1;
        eqn=ϕ[x]==f[x]+Integrate[Kernel[x,t]ϕ[t],{t,a,b}];
        ϕ[x]/.ApproxKernelByTaylor[Kernel,f,{a,b},ϕ,x,3]


Out[25]:= 1.0030774044049044 - 1.0030774044049044x+
        0.16736403228236804x³ + Sin[x]
```

Now, compare the exact solution $\varphi(x) = 1$ and approximate solutions for different n graphically (Fig. 5.5.9).

```
In [26]:= exactsol=1;
        nList=Range[3];
        approxsols=ϕ[x]/.ApproxKernelByTaylor[Kernel,
```

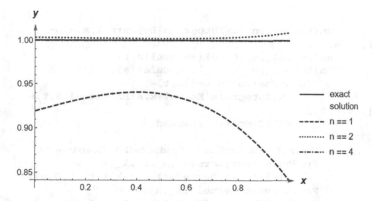

Fig. 5.9 Exact and approximate solutions of an integral equation obtained by replacing the kernel with a degenerate one (Example 5.12)

```
            f,{a,b},ϕ,x,#]&/@nList;
      legend=Append[ToString/@Thread[n==#]&@nList,
            "exact solution"];
      optplot={AxesStyle->Arrowheads[{0.0,0.025}],
            AxesLabel->{Style[x,Bold,Med-
dium],Style[y,Bold,Medium]},
            PlotRange->Automatic};
      Plot[{approxsols,exactsol}//Evaluate,{x,a,b},
            Evaluate@optplot,PlotLegends->legend]
```

The graphs of the approximate solution for $n = 3$ and the exact one practically coincide, which indicates a fairly good approximation.

Example 5.13 Solve the Fredholm equation of the second kind from Example 5.12.

$$\varphi(x) = \sin x + \int_0^1 (1 - x \cos xt)\varphi(t)\mathrm{d}t$$

replacing the kernel of the equation with a degenerate one and applying the method of Fredholm determinants.

Solution
We use the functions defined in Examples 3.1 and 3.2 to calculate the Fredholm determinants and to construct the resolvent.

```
In [1]:= Clear[Bn,Cn]
         Cn[Kernel_,0]=1;
```

```
        Cn[Kernel_,n_]:=Integrate[Bn[Kernel,x,x,n-
1],{x,a,b}]
        Bn[Kernel_,x_,t_,0]:=Kernel[x,t]
        Bn[Kernel_,x_,t_,n_]:=Module[{s},
            Cn[Kernel,n]Kernel[x,t]-
                n Integrate[Kernel[x,s]Bn[Kernel,s,t,n-
1],{s,a,b}]//
            FullSimplify//Expand
 ]
 In[6]:= Clear[FredholmMinor,FredholmDet,Resolvent]
        FredholmMinor[Kernel_,x_,t_,λ_,n_]:=
        Sum[(-1)^k/k! Bn[Kernel,x,t,k]λ^k,{k,0,n}]
        FredholmDet[Kernel_,λ_,n_]:=
        Sum[(-1)^k/k! Cn[Kernel,k]λ^k,{k,0,n}]
        Resolvent[Kernel_,x_,t_,λ_,n_]:=
        FredholmMinor[Kernel,x,t,λ,n]/FredholmDet[Ker-
nel,λ,n] //
            FullSimplify//Expand
```

Set the initial data and expand the kernel in the powers of t up to t^3.

```
 In [10]:= Clear[Kernel,f,a,b,λ,φ,x,n]
        Kernel=Function[{x,t},1-x Cos[x t]];
        f=Function[x,Sin[x]];
        a=0;
        b=1;
        λ=1;
        L=Function[{x,t},Evaluate@Normal@
            Series[Kernel[x,t],{t,0,3}]]
```
$$\text{Out[16]} := \mathbf{Function}\left[\{x,t\}, 1 - x + \frac{t^2 x^3}{2}\right]$$

Calculate the resolvent and construct a solution.

```
 In [17]:= Column@Table[{n,Cn[L,n],Bn[L,x,t,n]},{n,0,5}]
```
$$\text{Out[17]} := \begin{array}{c} \left\{0, 1, 1 - x + \frac{t^2 x^3}{2}\right\} \\ \left\{1, \frac{7}{12}, \frac{1}{12} - \frac{t^2}{8} - \frac{x}{12} + \frac{t^2 x}{8} - \frac{x^3}{24} + \frac{t^2 x^3}{4}\right\} \\ \left\{2, \frac{1}{16}, 0\right\} \\ \{3, 0, 0\} \\ \{4, 0, 0\} \\ \{5, 0, 0\} \end{array}$$

```
 In [18]:= res=Resolvent[L,x,t,λ,2]
```

$$\text{Out[18]} := \frac{88}{43} + \frac{12t^2}{43} - \frac{88x}{43} - \frac{12t^2x}{43} + \frac{4x^3}{43} + \frac{24t^2x^3}{43}$$

```
In [19]:= f[x]+λ Integrate[res f[t],{t,a,b}]//
          FullSimplify//Expand //N//Evaluate]
```

$$\text{Out[19]} := 1.0030774044049044 - 1.0030774044049044x +$$
$$0.16736403228236807x^3 + \text{Sin}[x]$$

As expected, the approximate solution completely coincides with the answer in Example 5.12.

References

S. G. Mikhlin, and K. L. Smolitskiy, *Approximate Methods for Solution of Differential and Integral Equations*, (American Elsevier Publ. Co., New York, 1967)

K.E Atkinson, The Numerical Solution of Integral Equations of the Second Kind. (– Cambridge: Cambridge Univ. Press, 1997)

H. Bateman, On the numerical solution of linear integral equations // Proc.Roy.Soc.(A). Vol.100. No. 705, 1922

L.M Delves, J.L. Mohamed, *Computational Methods for Integral Equations*. (- Cambridge – New York: Cambridge Univ. Press, 1985)

A. Golberg, Ed. Numerical Solution of Integral Equations. – New York: Plenum Press, 1990

W. Hackbusch, *Integral Equations: Theory and Numerical Treatment*. (- Boston: Birkhäuser Verlag, 1995)

M. L Krasnov, A. I. Kiselev, and G. I. Makarenko, *Problems and Exercises in Integral Equations*, (Mir Publ., Moscow, 1971)

Wolfram Mathematica, http://www.wolfram.com/mathematica/

Chapter 6
Individual Tasks. Passing the Final Test After Completing the Course

This chapter offers options for individual assignments for passing the final test after completing the course. In total, ten equivalent variants have been designed, each containing five tasks with the same level of complexity according to Bloom's Taxonomy.

With full-time study, these tasks can be useful for the students when consolidating the materials and to the teachers when preparing several options for test papers. The teachers can choose two or three tasks from the proposed list in the individual assignment to be implemented in the *Wolfram Mathematica* [1].

When working remotely in Sakai's distance learning environment, these individual tasks can be offered as a test. All tasks have the same level of difficulty. Therefore, during computer testing, a random selection of problems from the 50 tasks proposed below is allowed.

The proposed tasks are companions to the tasks in [2].

Assignment 1

1.1. Solve an integro-differential equation using the Laplace transform:

$$\varphi'(x) = x + \int_0^x \cos(x - t)\varphi(t)dt, \; \varphi(x) = 1$$

Answer: $\varphi(x) = 1 + x^2 + \frac{x^4}{4}..$

1.2. Solve the Fredholm equation with a degenerate kernel or prove that it has no solution

$$\varphi(x) = \cos x + \int_0^\pi tg \; x \cos t\varphi(t)dt$$

Answer: $\varphi(x) = \cos x - \frac{\pi}{2}tg \; x.$

© The Author(s), under exclusive license to Springer Nature Singapore Pte Ltd. 2022
V. Ryzhov et al., *Modern Methods in Mathematical Physics*,
https://doi.org/10.1007/978-981-19-4915-9_6

1.3. Solve the Volterra equation by reducing it to an ordinary differential equation:

$$\varphi(x) = e^x + \int\limits_0^x \frac{t}{t+1}\varphi(t)dt$$

Answer: $\varphi(x) = e^x\left(1 + x + \frac{x^2}{2}\right)$.

1.4. Investigate the solution of the Fredholm integral equation for various values of parameter λ:

$$\varphi(x) = 1 + \lambda \int\limits_0^\pi \cos(x + t)\varphi(t)dt$$

Answer: $\lambda \neq \pm\frac{2}{\pi} \rightarrow \varphi(x) = 1 - \frac{4\lambda}{2+\lambda\pi}\sin x$, $\lambda = \frac{2}{\pi} \rightarrow \varphi(x) = 1 - \sin x + C\cos x$, $\lambda = -\frac{2}{\pi} \rightarrow$ no solutions.

1.5. Solve the Volterra equation with a degenerate kernel:

$$\varphi(x) = 1 + \int\limits_e^x \frac{2}{t\ln x}\varphi(t)dt$$

Answer: $\varphi(x) = 2\ln x - 1$.

Assignment 2

2.1. Solve the Fredholm equation with resolvent using iterated kernels

$$\varphi(x) = e^x + \lambda \int\limits_0^1 e^{x-t}\varphi(t)dt$$

and indicate the region of convergence of the Neumann series. Find the solution of the integral equation for the specified value of the parameter λ and check the solution by direct substitution.
     Answer: $\varphi(x) = -e^x$.

2.2. Solve the Fredholm equation with a degenerate kernel or prove that it has no solution

$$\varphi(x) = e^{-x} + \int\limits_0^1 e^x t\,\varphi(t)dt$$

Answer: no solution.

2.3.   Solve the Fredholm equation using method of successive approximations

$$\varphi(x) = e^x + \frac{1}{2} \int_0^1 e^{x-t} \varphi(t) dt$$

Answer: $\varphi(x) = 2e^x$.

2.4.   Solve the Volterra equation by reducing it to an ordinary differential equation

$$\varphi(x) = 2sh\, x + \int_0^x (x - t)\varphi(t) dt$$

Answer: $\varphi(x) = xchx + shx$.

2.5.   Solve the Volterra equation with a difference kernel using the Laplace transform

$$\varphi(x) = e^{2x} - 2 + \int_0^x e^{x-t} \varphi(t) dt$$

Answer: $\varphi(x) = xe^{2x} + 1$.

Assignment 3

3.1   Solve an integro-differential equation using the Laplace transform

$$\varphi'(x) + \int_0^x e^{-2(x-t)} \varphi(t) dt = 0, \varphi(0) = 1$$

Answer: $\varphi(x) = e^{-x}(1 + x)$.

3.2   Solve the Fredholm equation with a degenerate kernel or prove that it has no solution

$$\varphi(x) = 5x + \int_0^1 \sqrt{xt} \varphi(t) dt$$

Answer: $\varphi(x) = 5x + 4\sqrt{x}$.

3.3.   Solve the Volterra equation reducing to an ordinary differential equation

$$\varphi(x) = 3\cos x - 4 \int_0^x (x - t)\varphi(t) dt$$

Answer: $\varphi(x) = 4\cos 2x - \cos x$.

3.4. Solve the Volterra equation with a degenerate kernel

$$\varphi(x) = x^2 + \lambda \int\limits_1^x \frac{2t}{x^2}\varphi(t)dt$$

Answer: $\varphi(x) = 2x^2 - 1$.

3.5. Solve the Volterra equation of the first kind using the Laplace transform

$$\int\limits_0^x \cos(x - t)\varphi(t)dt = \sin x - 2x$$

Answer: $\varphi(x) = 1 - x^2$.

Assignment 4

4.1. Solve the Fredholm equation with resolvent using iterated kernels

$$\varphi(x) = e^x + \lambda \int\limits_0^1 xe^{x-t}\varphi(t)dt, \lambda = -2$$

and indicate the region of convergence of the Neumann series. Find the solution of the integral equation for the specified value of the parameter λ and check the solution by direct substitution.

Answer: $\varphi(x) = e^x(1 - x)$.

4.2. Solve the Fredholm equation with a degenerate kernel or prove that it has no solution

$$\varphi(x) = 3\ln x + 2 \int\limits_0^1 \frac{x}{t}\varphi(t)dt$$

Answer: $\varphi(x) = 3\ln x - 2x$.

4.3. Solve the Fredholm equation using method of successive approximations

$$\varphi(x) = e^x + \int\limits_0^1 xe^{x-t}\varphi(t)dt$$

Answer: $\varphi(x) = e^x(1 + 2x)$.

4.4. Solve the Volterra equation by reducing it to an ordinary differential equation

$$\varphi(x) = \lambda \int\limits_1^x \frac{4t - 5x}{t^2} \varphi(t)dt + \ln x$$

Answer: $\varphi(x) = \cos 2 \ln x + \sin 2 \ln x - 1$.

4.5. Solve the Volterra equation using the Laplace transform

$$\int\limits_0^x ch(x - t)\varphi(t)dt = 3x^2$$

Answer: $\varphi(x) = 6x - x^3$.

Assignment 5

5.1. Solve an integro-differential equation using the Laplace transform

$$\varphi''(x) + \int\limits_0^x e^{2(x-t)}\varphi'(t)dt = e^{2x}, \quad \varphi(x0) = 0, \quad \varphi'(0) = 0$$

Answer: $\varphi(x) = e^x - 1$.

5.2. Solve the Fredholm equation with a degenerate kernel or prove that it has no solution

$$\varphi(x) = 2x - 1 - \frac{\pi}{2} \int\limits_0^1 (\cos \pi x - \sin \pi t)\varphi(t)dt$$

Answer: $\varphi(x) = C \cos x + 2x - 1 - \frac{2C}{\pi}$.

5.3. Solve the Volterra equation by reducing it to an ordinary differential equation

$$\varphi(x) = e^{2x} - 2x^2 - 2x - 1 + \int\limits_0^x \left[3(x - t) - (x - t)^2\right]\varphi(t)dt$$

Answer: $\varphi(x) = 4sh2x$.

5.4. Solve the equation with a symmetric kernel using the method of successive approximations

$$\varphi(x) = \sqrt{1 - x^2} + \int\limits_0^1 xt\varphi(t)dt$$

Answer: $\varphi(x) = \sqrt{1 - x^2} + \frac{x}{2}$.

5.5. Solve the Volterra equation of the first kind using the Laplace transform

$$\int\limits_0^x \varphi(t)dt = x^3 e^x$$

Answer: $\varphi(x) = (2+x)x^2 e^x$.

Assignment 6

6.1. Solve the Fredholm equation with resolvent using iterated kernels

$$\varphi(x) = \sqrt{1-x^2} + \lambda \int\limits_0^1 xt\varphi(t)dt, \lambda = 6$$

and indicate the region of convergence of the Neumann series. Find the solution of the integral equation for the specified value of the parameter λ and check the solution by direct substitution.
Answer: $\varphi(x) = \sqrt{1-x^2} + 2x$.

6.2. Solve the Fredholm equation with a degenerate kernel or prove that it has no solution

$$\varphi(x) = x + 2 \int\limits_0^{\frac{\pi}{2}} \cos(x-t)\varphi(t)dt$$

Answer: $\varphi(x) = x - 2\cos x$.

6.3. Solve the equation using the method of successive approximations

$$\varphi(x) = x^2 + \int\limits_0^1 \varphi(t)dt$$

Answer: $\varphi(x) = 2(e^x - x - 1)$.

6.4. Solve the Volterra equation by reducing it to an ordinary differential equation

$$\varphi(x) = 4x \ln x - 1 + \int\limits_1^x \frac{4x - 3t}{t^2}\varphi(t)dt$$

Answer: $\varphi(x) = x^3 - x - \frac{1}{x}$.

6.5. Solve the Volterra equation using the Laplace transform

$$\int\limits_{0}^{x} \varphi(t)dt = e^{2x}\sin x$$

Answer: $\varphi(x) = e^{2x}(\cos x + 2\sin x)$.

Assignment 7

7.1. Solve an integro-differential equation using the Laplace transform

$$\varphi'^{(x)} - \varphi(x) + \int\limits_{0}^{x}(x-t)\varphi'^{(t)}dt - \int\limits_{0}^{x}\varphi(t)dt = x, \varphi(0) = -1$$

Answer: $\varphi(x) = -e^x$.

7.2. Solve the Fredholm equation with a degenerate kernel or prove that it has no solution

$$\varphi(x) = 2 - 3\int\limits_{0}^{\frac{\pi}{2}}\sin(x - 2t)\varphi(t)dt$$

Answer: $\varphi(x) = 2 - 3\sin x$.

7.3. Solve the Volterra equation by reducing it to an ordinary differential equation

$$\varphi(x) = x^2 + \int\limits_{1}^{x}\frac{x}{t^2}\varphi(t)dt$$

Answer: $\varphi(x) = x^2(\ln x + 1)$.

7.4. Investigate the solution of the Fredholm integral equation for various values of parameter λ

$$\varphi(x) = \sin \pi x + \lambda\int\limits_{-1}^{1}(1 + xt)\varphi(t)dt$$

Answer: $\lambda \neq \frac{1}{2}, \frac{3}{2} \rightarrow \varphi(x) = \sin \pi x + \frac{2\lambda x}{\pi(1 - \frac{2\lambda}{3})}, \lambda = \frac{1}{2} \rightarrow \varphi(x) = \sin \pi x + \frac{3}{2\pi}x, \lambda = \frac{3}{2}$ no solutions.

7.5. Solve the Volterra equation using the Laplace transform

$$\varphi(x) = x^2 + \int\limits_{0}^{x}\sin(x - t)\varphi(t)dt$$

Answer: $\varphi(x) = x^2 + \frac{x^4}{12}$.

Assignment 8

8.1. Solve the Fredholm equation with resolvent using iterated kernels

$$\varphi(x) = \sin x + \lambda \int\limits_0^{\frac{\pi}{2}} x \sin t \varphi(t) dt, \lambda = 4$$

and indicate the region of convergence of the Neumann series. Find the solution
of the integral equation for the specified value of the parameter λ and check
the solution by direct substitution.

Answer: $\varphi(x) = \sin x - \frac{\pi x}{3}$.

8.2. Solve the Fredholm equation with a degenerate kernel or prove that it has no
solution

$$\varphi(x) = e^x + \int\limits_0^1 \left(e^x t + x e^t\right) \varphi(t) dt$$

Answer: $\varphi(x) = -3x$.

8.3. Solve the Fredholm equation using method of successive approximations

$$\varphi(x) = \sin x + \frac{1}{2} \int\limits_0^{\frac{\pi}{2}} x \sin t \varphi(t) dt$$

Answer: $\varphi(x) = \sin x + \frac{\pi x}{4}$.

8.4. Solve the Volterra equation by reducing it to an ordinary differential equation

$$\varphi(x) = \int\limits_0^x \cos(x - t) \varphi(t) dt + x$$

Answer: $\varphi(x) = 2e^x(x - 1) + x + 2$.

8.5. Solve the Volterra equation with a difference kernel using the Laplace transform

$$\varphi(x) = e^x + 2 \int\limits_0^x \cos(x - t) \varphi(t) dt$$

Answer: $\varphi(x) = \frac{e^x}{2}(x^2 + 4x + 2)$.

Assignment 9

9.1. Solve an integro-differential equation using the Laplace transform

$$\varphi''(x) + 2\varphi'(x) + \varphi(x) = \cos x + \int_0^x (x-t)\varphi''(t)dt + 2\int_0^x \sin(x-t)\varphi'(t)dt$$

Answer: $\varphi(x) = 1 - (1+x)e^{-x}$.

9.2. Solve the Fredholm equation with a degenerate kernel or prove that it has no solution

$$\varphi(x) = x^2 - 2\int_0^1 (3xt - 1)\varphi(t)dt$$

Answer: no solutions.

9.3. Solve the Volterra equation by reducing it to an ordinary differential equation

$$\varphi(x) = 6\int_0^x \cos 5(x-t)\varphi(t)dt - 4e^{5x}$$

Answer: $\varphi(x) = 10e^{5x} - e^{3x}(14\cos 4x + 13\sin 4x)$.

9.4. Solve the Fredholm equation using method of successive approximations

$$\varphi(x) = \frac{1}{2\pi}\int_0^\pi t\sin x\varphi(t)dt + \cos x$$

Answer: $\varphi(x) = \cos x - \frac{2}{\pi}\sin x$.

9.5. Solve the Volterra equation using the Laplace transform

$$\varphi(x) = \sin x + 3\int_0^x \sin 4(x-t)\varphi(t)dt$$

Answer: $\varphi(x) = 5\sin x - 2\cos x$.

Assignment 10

10.1. Solve the Fredholm equation with resolvent using iterated kernels

$$\varphi(x) = \ln x + \lambda \int_1^e \frac{\ln t}{x}\varphi(t)dt, \ \lambda = e$$

and indicate the region of convergence of the Neumann series. Find the solution of the integral equation for the specified value of the parameter λ and check the solution by direct substitution.

Answer: $\varphi(x) = \ln x - \frac{2e}{x}$.

10.2. Solve the Fredholm equation with a degenerate kernel or prove that it has no solution

$$\varphi(x) = 8x^2 - 5x + \int_0^1 (3x + 2t)\varphi(t)dt$$

Answer: $\varphi(x) = 4x(2 - x)$.

10.3. Solve the Fredholm equation using method of successive approximations

$$\varphi(x) = \ln x + \int_1^e \frac{\ln t}{x}\varphi(t)dt$$

Answer: $\varphi(x) = \frac{2e-4}{x} + \ln x$.

10.4. Solve the Volterra equation by reducing it to an ordinary differential equation

$$\varphi(x) = 2\int_0^x \sin(x - t)\varphi(t)dt + e^x$$

Answer: $\varphi(x) = chx + xe^x$.

10.5. Solve the Volterra equation using the Laplace transform

$$\varphi(x) = chx + 8\int_0^x sh(x - t)\varphi(t)dt$$

Answer: $\varphi(x) = ch3x$.

References

M. L. Krasnov, A. I. Kiselev, and G. I Makarenko, *Problems and Exercises in Integral Equations*, (Mir Publ., Moscow, 1971)
Wolfram Mathematica, http://www.wolfram.com/mathematica/

Printed in the United States
by Baker & Taylor Publisher Services